Flying the Big Jets

Flying the Big Jets

(All you wanted to know about the Jumbos but couldn't find a pilot to ask)

Stanley Stewart

Airlife
ENGLAND

ISBN 0 906393 69 8

First Published in 1984 by Airlife Publishing Ltd.
This Revised Edition Published 1986.

Printed in England by Livesey Ltd., Shrewsbury.

Airlife Publishing Ltd.

7 St. John's Hill, Shrewsbury, England.

Contents

Flying the Big Jets is dedicated to all aircrew, whether senior captains or junior flight attendants, but in particular to my colleagues on the flight decks of the world, any one of whom, had they flown the big jets and found the time, could have written this book and also dedicated to my sister Dorothy who died after a long illness and who was so much help in the preparation of this book.

Acknowledgements

I would like to thank Andrea Thomas for her invaluable help in correcting (and recorrecting) my English; the artist, Peter Bamber, for his wonderful drawings, and the photographer, Peter Pugh, for his excellent work in producing the prints. A special word of thanks must go to Mike Chan-Choong, 747 pilot, and Simon Robinson, 747 flight engineer, for their help with the manuscript. Any errors remaining are, of course, entirely my own.

Introduction

A myriad of books have been written on the subject of flying, from tales of the early attempts, right through the ages of flight, to space travel and science fiction fantasy. The two world wars inspired the pens of many a fine aviator, and instruction books abound on the principles of flight. Novels have been published on everything from airship disasters to Concorde dramas, the airline pilot's story has been told, and much has been printed on the airline world. So why another book on flying? Quite simply, because it is needed! The demands of an inquisitive public have now outstripped the material available, and the information presented in this book is designed to fill the gap. 'What is it really like to fly the big jets?' is a question that almost everyone seems to ask, and the one that this book hopes to answer.

A great many people have now flown, and those who haven't have seen enough on television or in the cinema to know a little of the airline world. They've a fairly good idea of what's involved for the passengers in going from, say, Paris to New York, and it takes only a little more information to fill in the basic procedures for the crew.

The crew arrive about one hour before departure, check the paper work and the weather, and a decision is made on the quantity of fuel required. The aircraft is then boarded and the pre-flight checks commenced. After the checks are completed, the passengers boarded, and the departure procedures studied, the engines are started. In radio contact with various controllers, and under their instructions, the aircraft taxies out, takes off, and sets course for its destination.

En route, the aircraft is guided along a predetermined track, passing from one radio control centre to another as the flight progresses. Approaching the destination, the arrival procedures are studied and, once again in liaison with a series of controllers, the aircraft starts the descent, completes the approach, let-down, and landing phases of the flight, and taxies to the terminal building. The engines are then shut down, and the final checks completed. After a long flight, the crew will go off duty, but after a short journey they may well be going on to another destination, so the pre-flight checks are begun once again and the whole procedure repeated.

That, of course, is all very simplified, but it covers loosely the basic procedures in flying an aircraft from A to B, and perhaps on to C. However, what on the surface appears to be a fairly straightforward procedure is in fact a complex operation. The flight crew require a great deal of training, knowledge, and skill to perform their tasks safely in what is potentially a hostile environment, notoriously unforgiving of error. Although most flights are routine, with so many lives at stake alertness and vigilance become second nature, and Murphy's Law probably applies more to the flying of aircraft than to any other task.

Murphy's Law states that: (1) nothing is as easy as it looks; (2) everything always takes longer than expected; (3) if anything can go wrong, it will — and at the worst possible moment.

That so few incidents do occur is due in no small measure to the respect afforded to Rule 3 by everyone concerned in aviation.

In the last few years much interest has been generated about the world of the big jets, and today the air travelling public is more than ever aware of its surroundings. The little information gained from a flight, or from watching aircraft at an airport or on film, is enough to whet the appetites of most for further knowledge. And what people want to know are the facts. They want to know the basic details of the flight. Any airline pilot knows the problems of being bombarded with questions in non-flying company once his occupation has been discovered. How often are the tyres changed? Does a pilot fly the same route all the time? Does he fly more than one aircraft type? Does he watch all the instruments at the same time? And a thousand similar questions.

To some the airline world is filled with magic and mystery where even the laws of nature are defied, and for a few the flying environment distorts imaginations and confuses even alert minds. It is not unusual for crews boarding the first stage of a long flight, say from Europe to Australia, or from the United States to the Persian Gulf, to receive parting comments from passengers that they'll see them again when they deplane in Sydney or Bahrain, some thirty hours later!

To be fair, however, the subject of flight holds many traps for the unwary because much is unexpected and the obvious often quite incorrect. Take one look at the Puffin bird with its over-large beak and odd shaped body and two facts become readily apparent — walking is achieved only with the greatest of difficulty and flying is impossible. No one, of course, told the Puffin bird! Ungainly in the air as it might be the Puffin most certainly does fly. Aircraft, however, although extremely complicated, are pieces of mechanical and electrical equipment just like a sewing machine or a locomotive, and need to be looked after, oiled and maintained in exactly the same way. All airlines, for example, instead of using new tyres to replace old ones, use retreads wherever possible, just like on the family car; a fact that seems to amaze everyone who hears it! The Boeing 747 also has quite recognisable windscreen wipers and washers! Crews too, in general, are fairly ordinary, straightforward people, doing a job of work just like anyone else, with many of the same interests, but with, perhaps, a few specialized problems of their own. It has not been unknown for a pilot suffering from, say, a sprained ankle, to be told at a hospital casualty department that he'll be back to work in no time, the staff little realizing that the rudder, one of the basic flying surfaces, is controlled by pedals, and the brakes operated by pressure from the toes. Even a slight loss of strength or movement in a foot could prove disastrous! With misunderstanding of this nature it's not surprising that most airlines employ their own specialized medical personnel.

Within these pages as many questions as possible have been answered, and much information has been added on the training, knowledge and skills of pilots, together with facts and figures to enlighten and amuse the reader. *Flying the Big Jets* doesn't attempt to tell a story, but merely presents the information that people want to know in a plain and simple manner. Although much of the material is of a technical nature the book is not a technical manual but an elementary introduction to airline flying written specifically for the layman with an interest in the big jets. Explanations are given so as to be understood by all with a very basic understanding of the sciences, with the drawings and photographs being added where required.

The book has been written 'through the eyes of a pilot', and in 'The Facts' much detail is given to prepare the reader for the 'pilot's seat' on an imaginary trip in 'The Flight'. Since the range of general aviation material is large much has been omitted in concentrating on the big jets, but care has been taken not to treat important subjects lightly. In understanding the big jets a certain basic aviation knowledge is required, but the reader is taken from the basics to the big jets in one easy leap. In a book of this nature some subject overlap is inevitable, as flight itself is the result of so many different interrelating factors, but repetition of detail has been kept to a minimum. Where required, references have been included in brackets when cross-referring to information in other sections.

Aviation language is full of abbreviations to which it is necessary to introduce the reader; for example ADF, VOR, DME, and so on.

However, extensive use of unfamiliar abbreviations is tiresome, and has been avoided where possible. To prevent confusion and aid the reader's memory, fully expanded terms, with abbreviations in brackets, have been repeated at regular intervals, e.g. automatic direction finder (ADF). A list of abbreviations is also included at the end of the book.

It is hoped that this book will meet at least some of the demands of those seeking further information, but, of course, it will not satisfy all. To begin with, airline pilots fly mostly only one aircraft type, for example the Boeing 747, since the complexities of modern aeroplanes make it difficult for crews to fly more than one type at a time. Airliners vary greatly in construction and size, and what would be normal practice on one could be potentially dangerous on another. Also, pilots tend to be divided between long haul on world wide routes, and short haul on continental flights, and what is true for one group may not be true for the other. Airlines, too, sometimes operate quite differently from their competitors, even flying the same type of aircraft on the same routes.

The facts and figures presented here are derived mainly from the Boeing 747 and its operation as flown on a world wide basis, but although the book has been written as comprehensively as possible, it is obvious from what has been said that there are bound to be a few omissions and inconsistencies; crews will not always operate as stated, nor will pilots everywhere live their lives in such a manner. However, whenever a pilot is invited by non-flying friends to a dinner party, he can now take along several copies of *Flying the Big Jets,* distribute them beforehand, and thereby enjoy his meal in peace!

Part 1

The Facts

Chapter 1
Principles of Flight

An aircraft flying straight and level is influenced by four forces, as shown in Fig. 1.1, and is in balanced flight when they are in equilibrium, i.e. when lift equals weight and thrust equals drag.

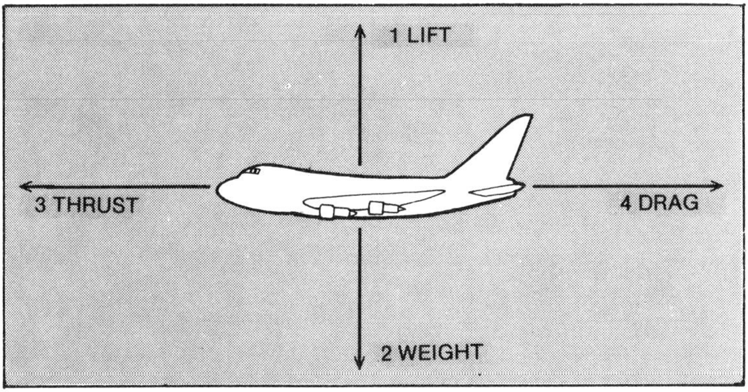

Fig. 1.1 The four forces acting on an aircraft.

1. **Lift** is the upward force created by the wings and is assumed to act through a central point known as the centre of pressure.
2. **Weight** of an aircraft is expressed in either kilograms or pounds and is assumed to act through a central point known as the centre of gravity.
3. **Thrust** is the force of the engines, normally expressed in pounds, which propels the aircraft forward through the air and is assumed to act in line with drag.
4. **Drag** is the result of the air resisting the motion of the aircraft.

Lift

If a driver extends his arm out of a moving vehicle and holds his flat hand inclined to the airflow, the flow of air passing over the surface of the hand produces a force that lifts the hand upwards and pushes it

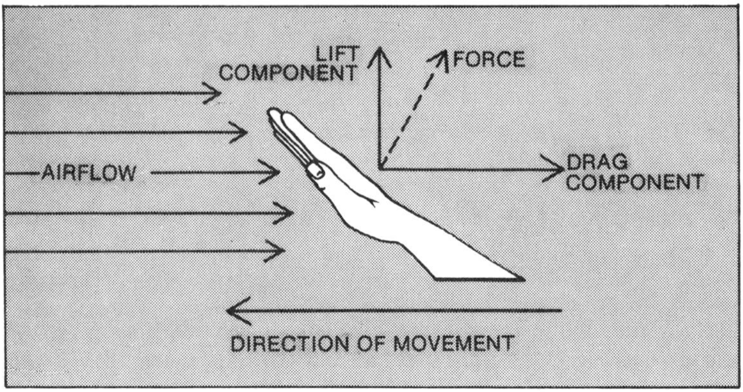

Fig. 1.2 Lift and drag.

backwards (Fig. 1.2). The upward component of the force is known as lift and the backward as drag. A wing is a more refined shape than a flat hand but produces lift in exactly the same way, although a lot more efficiently. An aircraft wing is fixed to the structure at an angle relative to the airflow as it flies through the sky. Air going the long way round up and over the curve of the wing is forced to increase speed resulting in an area of low pressure being induced on the top surface that draws the wing upwards. Some lift derives from the airflow striking the lower surface of the wing creating an increase in pressure forcing the wing upwards, but the greater lift results from the reduction in pressure above.

The area of low pressure on the top of the wing is not a vacuum but simply a reduced value of pressure relative to the surrounding air, and is shown as negative pressure. The area of high pressure below

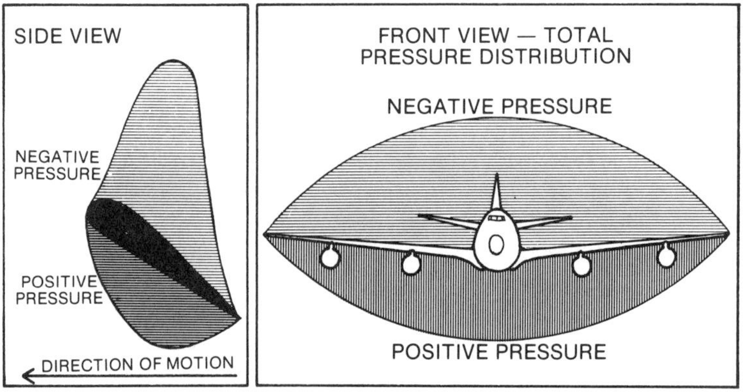

Fig. 1.3 Pressure pattern distribution around an aircraft.

the wing is similarly an increased value relative to the surrounding air and is shown as positive pressure. The pressure pattern distribution surrounding an aircraft (Fig. 1.3) clearly shows the greater effect of the negative pressure in the lifting process. To describe lift in more precise terms it can be said that the low and high pressure areas above and below the wing combine at the trailing edge as a downwash from which the wing experiences an upward and opposite reaction in the form of lift. Thinking of lift in simple terms, however, it is not so ridiculous as it seems to imagine the aircraft being sucked into the air by the reduced pressure above the wings.

Lift is affected by a number of factors. The density of the air affects lift: the higher the density the greater the lift. The airspeed over the wing, i.e. the true airspeed (TAS) of the aircraft, affects lift: the faster the speed the greater the lift. The angle at which the wing is inclined to the airflow, known as the angle of attack (Fig. 1.4), affects lift: the larger the angle the greater the lift. Since the wings are firmly fixed to

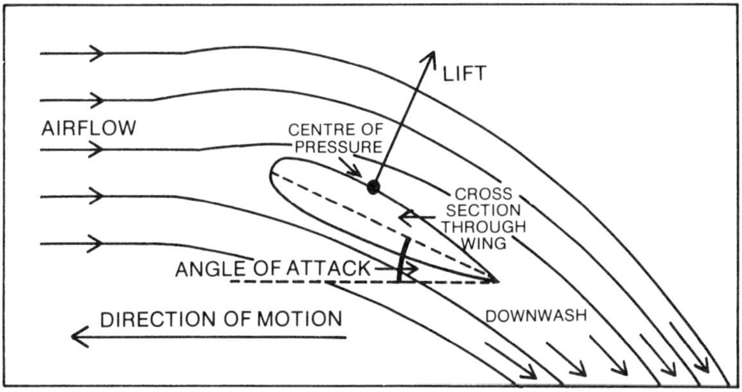

Fig. 1.4 Angle of attack.

the structure, the angle of attack is varied by pitching the aircraft nose up or down and is referred to as the attitude of the aircraft. To maintain constant lift, therefore, as in level flight, variation in true airspeed requires adjustment of aircraft attitude; i.e. faster airspeeds require a lower nose attitude and slower airspeeds a higher nose attitude. Wing surface area is also a function of lift: the larger the area, the greater the lift. The bigger and heavier the aircraft, therefore, the larger the wing span and wing surface area required to produce sufficient lift. Today's large jets are constructed with wings of enormous size, the Boeing 747 having a wing span of 60 metres (195 feet).

Leading edge flap fully set on approach.

On modern jets the wings are swept back at a large angle (37°) to allow aircraft to cruise at high speeds by delaying the onset of shock waves as the airflow over the wing approaches the speed of sound (*see* Flight Instruments p. 124). At slow aircraft speeds, however, the lift-producing qualities of the wing are poor. High lift-producing devices in the form of leading and trailing edge flaps are required and when extended increase the wing surface area and the camber of the wing shape (Fig. 1.5). With flaps fully extended the wing area is increased by 20 per cent and lift by over 80 per cent. Flaps increase lift allowing slower speeds and also increase drag which retards the aircraft. Flap drag can also be used to increase the rate of descent. Canoe-shaped fairings below the wings shroud the tracks and drive mechanisms used in flap lowering.

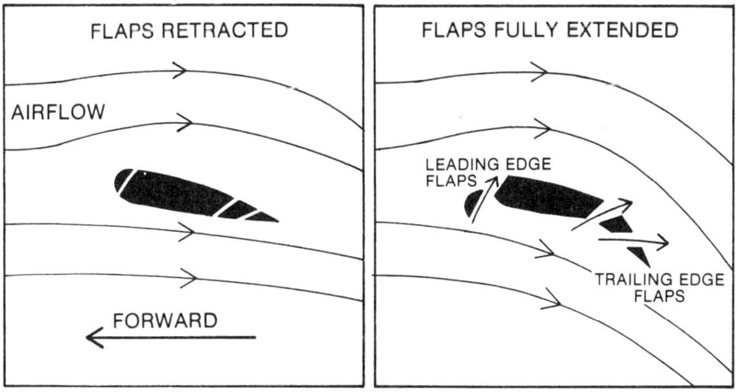

Fig. 1.5 The effect of flaps on wing surface area and camber.

To improve lift at take-off, flaps are set at 10 or 20 degrees, depending on circumstances, any increase in drag being more than compensated by increase in lift. Take-off without flap is not possible at normal operating weights. On landing, full 30 degree flap is always selected in normal circumstances.

Large jets departing fully laden on long haul flights require long take-off runs of the order of 50 seconds duration before becoming airborne. At the required speed for take-off the pilot raises the aircraft nose (called rotation) to a predetermined pitch angle to increase the angle of attack of the wing to the airflow with a resultant increase in lift, and the aircraft climbs into the air. At maximum take-off weight the Boeing 747 requires a take-off speed of 167 knots (192 mph or 309 km/hr), and major airport runway lengths are normally about 3½ kilometres (around 2 miles) to accommodate the take-off distances required. Not all take-offs, of course, are at maximum

Clean wing. Note the canoe-shape fairings below the wing which shroud the flap mechanism.

Trailing edge flap 20° set on approach to land.

weight, and at lower weights less lift is required. The aircraft lifts off at slower speeds and therefore requires a shorter run along the runway.

Weight

Although aircraft weights are normally given in kilograms or pounds, the enormous weight of today's jumbos becomes meaningless to many people when expressed in hundreds of thousands of a particular unit, and an appreciation of the weights involved is often better achieved by stating them in larger terms. One tonne (or metric ton) is equal to 1000 kg, which also equals 2200 lbs. One ton is equal to 2240 lbs. One tonne, therefore, is almost equivalent to one ton, being only 40 lbs lighter. Whether units are stated in metric or imperial, or pronounced as tonnes or tons, can be seen to make little difference, and to simplify matters all weights are expressed in tons. Take, for example, the maximum take-off weight of a Rolls-Royce engined Boeing 747 of 820,000 lbs. Stating this weight as 372 tons brings home to most the enormous size of the aircraft.

Aircraft loading

To the basic weight of an aircraft is added the weight of equipment and the weight of the crew and their bags, the resultant figure being

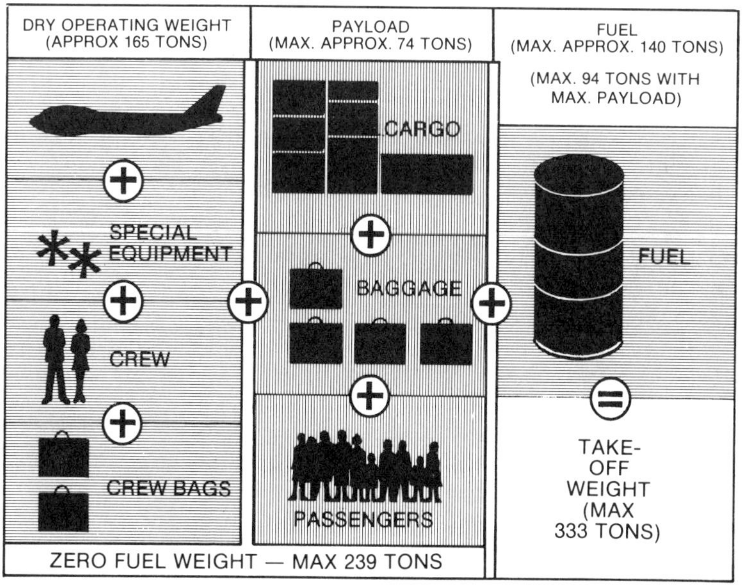

Fig. 1.6 Take-off weight.

Landing aircraft, gear down, flaps fully set.

known simply as the dry operating weight. To this weight is added the payload, which consists of the weight of the cargo (including passenger baggage) and the weight of the passengers (males at 78 kg/170 lbs, females at 68 kg/150 lbs, children at 43 kg/95 lbs and infants at 10 kg/22 lbs — including hand luggage and necessities). The dry operating weight and the payload account for all weights excluding fuel and together are known as the zero fuel weight. To the zero fuel weight is added the weight of the fuel to obtain the final take-off weight (Fig. 1.6). The total aircraft weight at any point in the flight is known as the all up weight (AUW).

The approximate Boeing 747-100 series dry operating weight is 165 tons. Since the maximum structural take-off weight is 333 tons, the maximum weight of 168 tons of payload and fuel able to be carried is more than the jumbo's own weight! The maximum fuel load depends on the specific gravity of the fuel and the maximum capacity of the fuel tanks and is about 140 tons, which is equivalent to the maximum take-off weight of a fully laden Boeing 707.

The maximum number of passengers depends on the maximum number of seats it is possible to fit, which, when coach seats are fitted in the upper deck lounge is 552. (Qantas holds the record for the greatest number of people carried on one flight. 674 were packed aboard a Boeing 747 and airlifted out of Darwin during the emergency evacuation after the cyclone struck in 1974). The normal passenger figures on a scheduled flight vary from about 400 to 450.

Trailing edge flap 30° set (landing flap).

Section of canoe shaped fairing and trailing edge flap drive mechanism exposed.

Average weights for a Boeing 747 on an eight-hour flight are: dry operating weight 165 tons; payload 35-40 tons; fuel 105-110 tons (of which about 90 tons is used, the rest in reserve); and take-off weight 305-315 tons. The maximum structural landing weight of the Boeing 747-100 series is 265 tons.

Weight and Balance

Weight distribution on an aircraft is very important: incorrect loading can result in the aircraft being too nose heavy or too tail heavy and beyond the ability of the controls to correct. Payload weights and distribution are, therefore, carefully pre-planned. Most cargo (including passenger baggage already weighed at check in) is pre-loaded on pallets designed to fit the shape of the hold. The pallets are raised to the level of the cargo door on special loading vehicles and slid on rollers from the raised platform into positions in the hold. The weight and position of each pallet is carefully noted. The required weight of fuel is decided by the Captain, and this weight converted to a volume by using the specific gravity of the fuel. It is then pumped aboard by the litre or gallon into tanks in the wings and belly of the aircraft.

Passenger weights and seat allocations are noted at check-in and fed to a computer, which also receives information on the cargo distribution and final fuel load. The computer then calculates the centre of gravity and checks that this is within limits. The aircraft is designed to cope with a range of movement of the centre of gravity to allow for take-off at different weights and with varying weight distribution. During flight the fuel weight distribution changes with fuel consumption, resulting in movement of the centre of gravity. The computer therefore also calculates that the centre of gravity remains within limits for the entire flight. All this information is noted on a load sheet that is presented to the Captain for inspection and signature when the final loading is completed just before departure.

The importance of accurate weight and balance calculations is illustrated by an incident that occurred to a big jet in Chicago in 1980. On take-off, acceleration along the runway was slower than expected and lift-off required more than the normal effort from the pilot. *En route* climb was poor and cruise at 35,000 feet required an increase in engine power to maintain a lower than normal airspeed. Descent and landing, however, were normal. On deplaning a passenger asked jokingly if the plane felt heavy, explaining that most of the passengers were coin collectors going to a convention and had brought coins on board in hand luggage. On investigation it was discovered that the hand luggage amounted to almost one and a half tons of unreported weight, and had the balance been critical a serious incident could have resulted.

Thrust

Static thrust is the thrust developed by a jet engine with the aircraft stationary and maximum take-off power set, and is stated in pounds. Since the performance of a jet engine is proportional to the density of the intake air, the aircraft is assumed to be at sea level in the standard atmosphere of 15°C (59°F) and pressure 1013.2 millibars (29.92 inches of mercury). The latest Rolls-Royce engines on the Boeing 747-200 series each develop 50,000 lbs of static thrust, giving a total of 200,000 lbs of thrust produced by the four engines in the take-off condition.

Drag

The two basic types of drag are profile drag, caused by the shape and skin surface of the aircraft, and induced drag, a side effect of the production of lift.

Profile drag

Drag produced by the shape of the aircraft is a result of the smooth flow of air being diverted round the form of the aircraft and is in fact know as form drag. The streamlined structure of an aircraft is designed to reduce form drag to a minimum.

Drag is also produced by friction between the aircraft skin surface and the airflow and this is known as skin friction. Air flowing over a surface results in a layer of retarded air being formed in immediate contact with the surface over which it is passing. (Water in a river, for example, always flows faster in the middle than at the banks due to the same effect). This retarded layer is known as the boundary layer and its thickness depends on the type of surface over which the air is flowing. Aircraft surfaces are highly polished to produce a thin boundary layer that maintains skin friction at a minimum.

Profile drag, then, is a combination of form drag and skin friction and is related to the speed of the aircraft, increasing markedly as the aircraft speed increases — doubling the speed of the aircraft quadruples the profile drag produced. (Any cyclist knows the problems of pedalling against a head wind as opposed to cycling in calm conditions.)

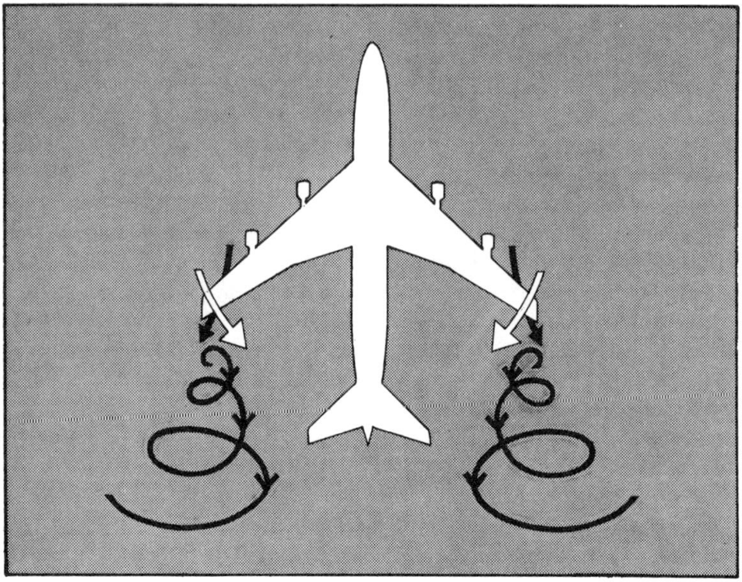

Fig. 1.7 Wing tip vortices.

Induced drag

Induced drag is a direct result of the production of lift and is caused by the mixing of the upper and lower airflows at the trailing edge of the wings. The airflow over the top surface of the wing tends to flow inwards towards the maximum low pressure area produced above the wing root, and the airflow under the wing tends to flow outwards from the maximum high pressure area produced below the wing root. The two airflows meet at an angle at the trailing edge of the wing and combine to produce a rotating airflow at each wing tip known as a wing tip vortex (Fig. 1.7). These wing tip vortices rotate in the direction of the wing root and result in a high level of turbulent airflow being produced in the wake of a large aircraft. The effect of speed on induced drag is quite different from profile drag in that induced drag actually decreases with an increase in airspeed. Wing tip vortices, therefore, are more evident at slow speeds during both take-off and landing, but are most pronounced on the final approach to landing. They can be clearly seen when watching aircraft land on a rainy day with a lot of moisture in the air.

Total drag

As has been stated, profile drag increases with *increase* in airspeed and induced drag increases with a *reduction* in airspeed. Total drag at

Downwash at the trailing edge of the wings and wing tip vortices can be clearly seen on this landing aircraft.

any one time therefore consists partly of profile drag and partly of induced drag. When total aircraft drag is plotted against speed by counting the effects of the two drag types, the graph shown in Fig. 1.8 results. The graph shape is known as the drag curve and is familiar to all pilots.

The speed for minimum drag on the total drag curve corresponds to the point where profile drag and induced drag are equal. At this

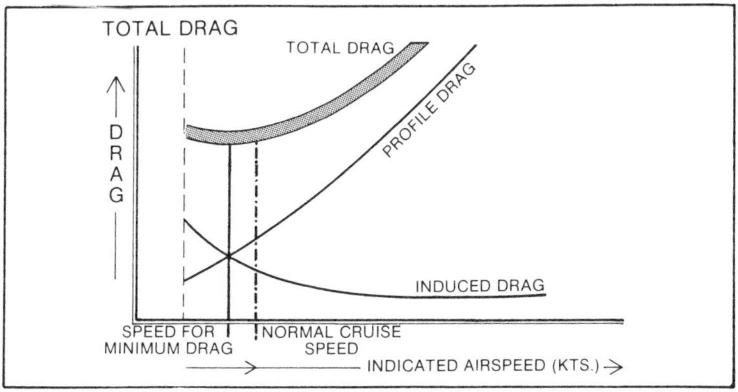

Fig. 1.8 Total drag curve.

point an uncanny situation arises whereby either an increase or a decrease in speed results in an *increase* in drag. Reduction of only a few knots from this minimum drag speed results in the aircraft entering the 'wrong' side of the drag curve where the drag *increases* rapidly with *reducing* airspeed, and large amounts of power are required to increase aircraft speed. Cruising speeds are normally in excess of the speed for minimum drag to maintain the aircraft on the 'right' side of the drag curve to ensure a safe operating margin.

Stalling

As an aircraft slows, to maintain lift the angle of attack of the wing is increased by raising the nose of the aircraft, with a resultant increase in induced drag. If the aircraft speed is allowed to become too slow and the nose up attitude required is excessive, a point is reached at which the angle of attack becomes critical and the smooth airflow over the wing breaks away from the upper surface producing turbulent flow (Fig. 1.9). With a breakdown in the smooth airflow all lift is lost and maximum drag results from the turbulent wake. This condition is known as stalling. With lift lost from the wings the aircraft nose pitches down (normally up on 'T'-tailed, rear-engined aircraft) and the aircraft flutters flatly from the sky like a falling leaf. The onset of the stall is accompanied by buffeting and shaking due to the turbulent wake produced. Recovery is achieved by forcing the aircraft into a dive by pushing the control column forward, and by applying full power until flying speed is once again achieved. The aircraft can then be pulled out of the dive and flown straight and level. Obviously, stalling on large jet aircraft is an extremely hazardous manoeuvre, and pilot training covers thoroughly the

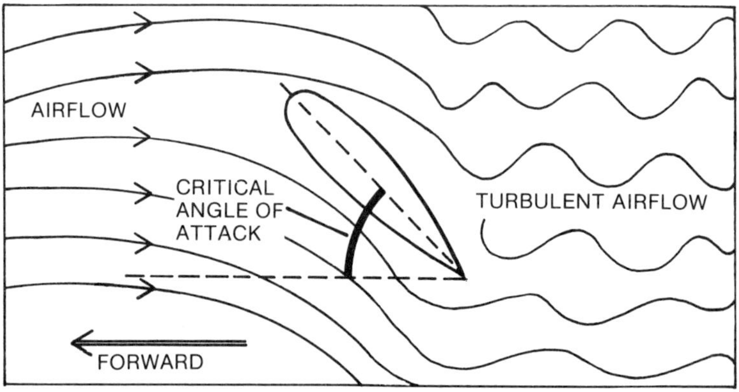

Fig. 1.9 Stalling-critical angle of attack.

recognition of and recovery from an early approach to the stall condition long before stalling actually occurs. Stall warning devices include a 'stick shaker', which physically shakes the control column at early onset of the stall condition, and a 'stick pusher', which delivers a hefty push forward to the control column at later stall development. Stalling speeds for the Boeing 747 at 250 tons (an average landing weight) are 165 knots (190 mph/305 km/hr) indicated airspeed clean (i.e. with no flaps or gear extended) and 110 knots (125 mph/205 km/hr) indicated with flaps fully down and gear lowered.

In-flight balance and stability

Balance

Imagine a model aircraft in a child's bedroom hanging from the ceiling by a thread. The aircraft is balanced when the thread is attached to the point through which the centre of gravity acts. If the thread is attached aft of the centre of gravity the nose pitches down, and if attached forward of the centre of gravity the nose pitches up. A similar effect results if the point of attachment of the thread remains fixed and the centre of gravity of the model is made to move forward or aft of this point.

Imagine now an aircraft in straight and level flight suspended by the upward lift force acting through the centre of pressure in a similar manner to the model aircraft held aloft by the thread acting through the point of attachment. In the air the centre of pressure acts like a pivot, similar to the central point of a see-saw, and to maintain the aircraft in balanced flight the weight is required to act in line with lift. Attempting to balance an aircraft in this position throughout flight, however, would be like trying to balance the model aircraft on a knife edge, and is quite impractical. Lift and weight seldom act in line, owing to movement of the centre of pressure with flap selection on take-off and landing, and to the rearward movement of the centre of gravity with fuel consumption in cruise. (Fuel is first used from the centre and inboard main wing tanks and last from wing tip tanks which are well aft due to the swept back wings — *see* Fuel p. 42.) The problem is overcome by the addition of a movable tailplane that acts as a stabilizing agent; the whole tailplane being designed to move until positioned to redress any imbalance. (This is not to be confused with the movement of the elevator which forms part of the tailplane and which is discussed later.) The variable position tailplane has been aptly named the horizontal stabilizer.

When the centre of gravity is aft of the centre of pressure the aircraft is tail heavy, and a further force is required to counteract the effect and stabilize flight. This is achieved by increasing the angle of attack of the stabilizer by hydro-mechanically moving the complete tailplane. Similarly, when the centre of gravity is forward of the centre of pressure and the aircraft is nose heavy, the angle of attack of the stabilizer is decreased by hydraulically moving the tailplane to produce negative lift, which acts downwards, and which once again counteracts the displacement of the forces and balances the aircraft. This process of balancing the aircraft by movement of the stabilizer is known as trimming. When handling the aircraft the out-of-balance forces can be felt by the pilot as pressure on the control column. Operation of electric switches on the control column activates the hydraulic mechanism which moves the stabilizer. As the aircraft is trimmed the control column is relieved of the out-of-balance pressure, and the aircraft is balanced when the control column is free of pressure forces and a stable aircraft condition is maintained with hands off the controls.

During take-off and landing the handling pilot has continually to retrim the aircraft as flight conditions change. In the cruise, when the autopilot is normally selected, trimming is controlled automatically.

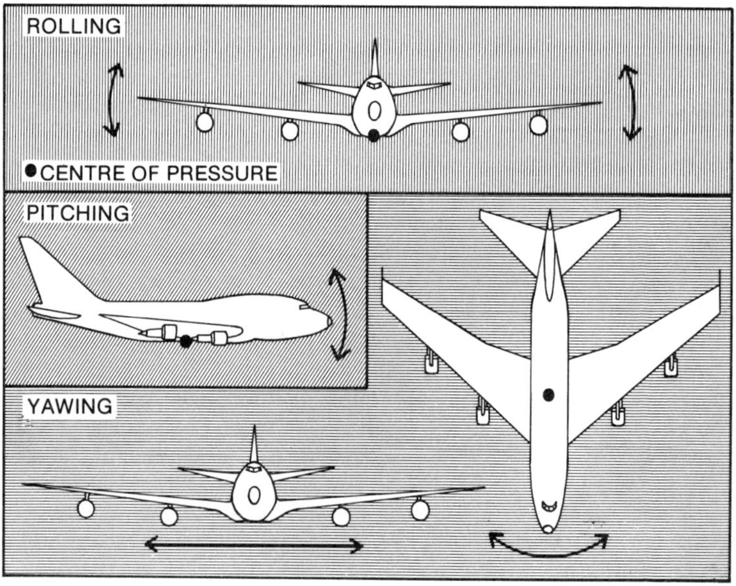

Fig. 1.10 Centre of pressure.

Before departure a computer calculates the stabilizer setting required to 'balance' the aircraft in the air just after take-off for the particular weight and load distribution concerned. The relevant stabilizer setting is then set on the trim scale during ground checks just before take-off.

Stability

If an object is displaced and returns to its original position it is said to be stable, and if it does not, unstable. Aircraft are designed with a degree of natural stability, and when disturbed from their original line of flight by a gust of wind, attempt to return to the initial stable flight condition without movement of the flying controls. In large passenger transport jets a good degree of stability is desirable, and inherent stability in the three aircraft movements of pitch, roll and yaw is a feature of basic aerodynamic and structural design.

Natural flight stability

A simple picture of each movement can be drawn by imagining the centre of pressure of the aircraft as a pivot point about which the aircraft moves in all directions (Fig. 1.10).

Stability in pitching motion is a function of the tailplane, just like the fins of a missile or flights of a dart. (Not to be confused with the movement of the stabilizer in trimming the aircraft to maintain balanced flight.) If the aircraft nose is pitched up by a gust of wind, the angle of attack of both the wings and the tailplane is increased. The extra lift produced by the tailplane, being far from the centre of pressure, is sufficient, due to the long leverage, to raise the tail and return the aircraft to straight and level flight. The opposite results with nose pitched down. (Fig. 1.11).

Stability in rolling motion is a function of the dihedral construction of the wings, i.e. each wing plane is positioned at a slight angle (7°) to the horizontal. If the aircraft is rolled by a gust of wind it slips to the side in the direction of roll. As the aircraft sideslips air resistance below the lower wing pushes the wing up, while the upper wing, positioned behind the aircraft body due to the dihedral effect, is protected from this sideways airflow, and the aircraft returns to level flight.

Stability in yawing motion is a function of the tailfin. If a gust of wind yaws the aircraft to the left or right the tailfin is momentarily displaced from its position. The resultant force, once again being far from the centre of pressure, is sufficient to return the tail to its original position.

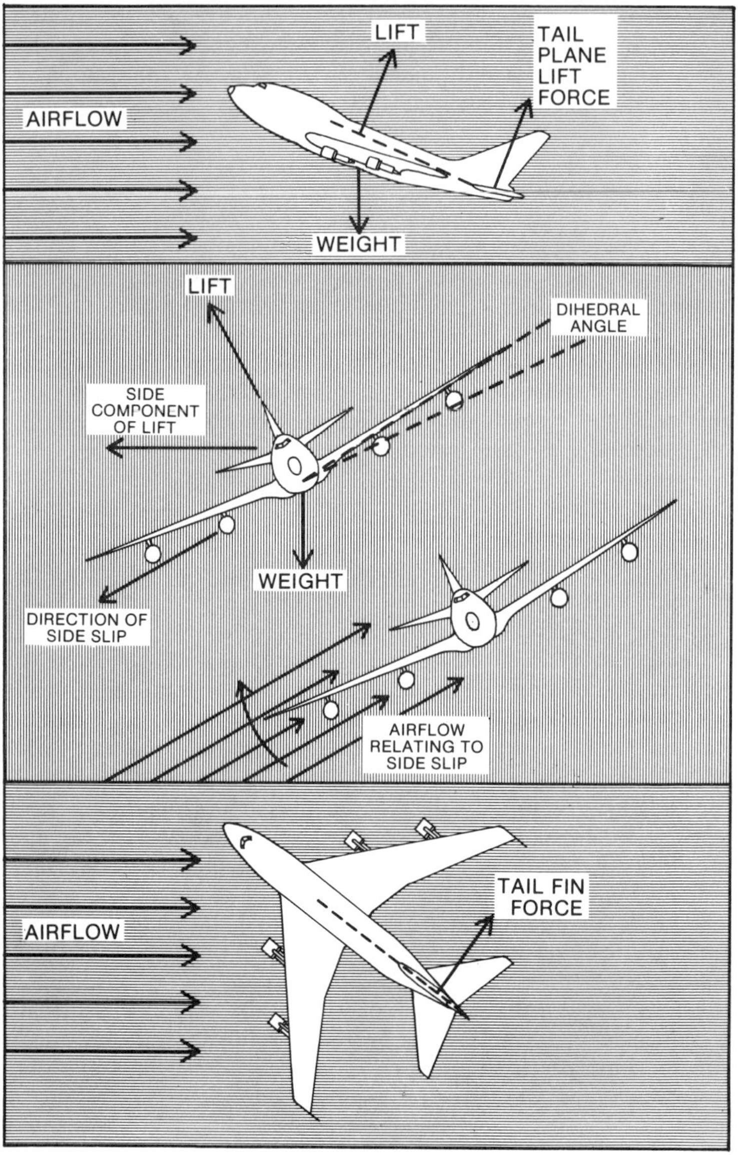

Fig. 1.11 Natural flight stability.

Flying control surfaces

The elevators control aircraft movement in pitch, ailerons in roll, and rudder in yaw (Fig. 1.12). All control surfaces are operated

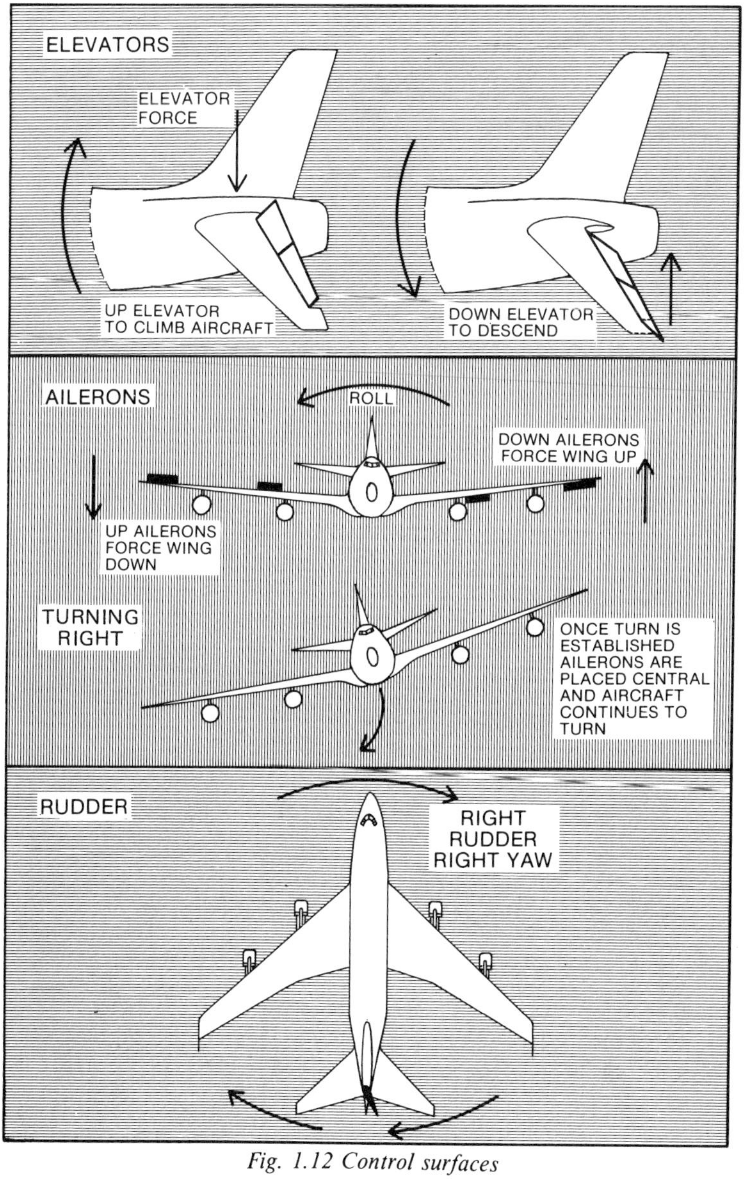

Fig. 1.12 Control surfaces

Slow speed left turn (flap 10° set) inboard and outboard ailerons both operating.

hydraulically and are split into sections, each powered by one or more of four separate hydraulic systems, thus minimizing any loss due to system failure. When functioning normally the control surface sections of the control unit being operated move in unison. Displacement of the control surfaces from the central position results in the airflow over the surface of the control applying a force that moves the aircraft in the required direction.

The elevators — climbing and descending

Upward movement of the elevator results in a negative lift force being applied that forces the tail down and, therefore, the nose up, and the aircraft climbs. Downward movement of the elevator results in descent.

The ailerons — rolling and turning

Turning to the left or right is achieved by the ailerons which roll the aircraft, resulting in a turn. Rolling of the aircraft in a turn is similar to the banking of a motorbike in a turn. (Turning is not the function of the rudder as it is on a ship.) One aileron set moves up, reducing lift, which forces the wing down, while the opposite set moves down, increasing lift, which forces that wing up. When the required bank is applied (the greater the bank the faster the turn) the ailerons are placed centrally and the aircraft continues to turn. Straightening of the aircraft is achieved by opposite application of the controls.

At high speeds only small aileron movements are required and operation of the outboard ailerons is inhibited. When good turning ability is required at low speeds (after take-off and on the approach) both sets of ailerons operate. The system is programmed by flap selection, and both aileron controls are active when flaps are selected down.

The rudder — yawing

The rudder is only used for directional guidance (i.e. like the rudder of a ship) when the aircraft is on the runway and is accelerating for take-off or decelerating after landing. While taxying a tiller is used to steer the nose wheel. The rudder is used during asymmetric flight (i.e. with an engine failure) to redress the imbalance caused by greater engine power on one side than on the other (*see* Flight Instruments p.120). At high speeds only small rudder movements are necessary and a rudder ratio system reduces rudder movement with increase in airspeed. The rudder incorporates a turn co-ordinator, which is programmed by selection of the flap to operate the rudder automatically in the direction of turn as an aid in turning the aircraft at slower speeds.

The rudder also acts as a yaw damper and operates automatically to suppress involuntary movement of the aircraft in roll and yaw, known as Dutch roll (so named from the inability of the early Dutch sailors to walk straight on land after many months at sea and much alcohol!). Dutch roll is usually initiated by a gust which results in a yawing-rolling oscillation due to the poor damping qualities of the swept back wing. A small gyro/computer senses the motion and signals the rudder to apply an opposing movement that dampens the Dutch roll.

Spoilers — speed brakes

Spoilers (Fig. 1.13) are so called because they spoil the lift of the wing by disrupting the airflow on the upper surface. On landing the spoilers automatically deploy to spoil the lift and place the full weight of the aircraft firmly on the wheels. This helps prevent the aircraft bouncing back into the air after a heavy landing and also improves braking capacity. On an abandoned take-off, selection of reverse thrust automatically deploys the spoilers, which once again places the full aircraft weight on the wheels to improve braking.

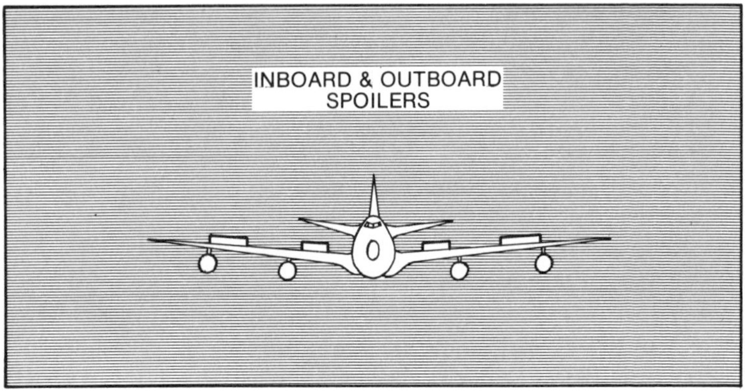

Fig. 1.13 Spoilers.

In flight the spoilers can be used as speed brakes, and are deployed by manual operation of a lever to slow down the aircraft rapidly or greatly increase the rate of descent. A gentle rumbling can be detected in the cabin when speed brakes are extended. Spoilers also operate as a flying control when a higher roll response is required by automatically deploying on one side to aid the aircraft during turns. Raising of the spoilers on the down wing side reduces lift which further forces down the down-going wing.

High speed left turn. Inboard aileron and spoilers operating.

Aircraft at touchdown. Full flap set, spoilers fully raised.

Spoilers extended as speed brakes (left wing similar).

Flying controls — Control column and rudder pedals

Movement of the controls (Fig. 1.14) is instinctive. On take-off, when flying speed is achieved, the control column is pulled back to 'rotate' the aircraft to the required nose up attitude for lift off. On landing, the control column is pulled back to 'flare' the aircraft to arrest the rate of descent for a smooth touch down. In flight, pulling back on the control column raises the nose and climbs the aircraft; pushing forward results in descent. Turning the yoke to the left (like the steering wheel of a vehicle) banks the aircraft to the left, with subsequent turning, and *vice versa*. Left rudder yaws the aircraft to the left and right rudder to the right. Above the rudder footrest on each side is a toe brake, which operates the brake by pressure from the toes. The toe brakes apply braking to the wheel bogies on their respective sides allowing differential braking (like directional control on a tank — braking the right track turns the tank to the right, etc.) to supplement rudder control if required on deceleration after landing or on abandoned take-off.

Direct physical movement of the control surfaces on large jet aircraft is beyond human strength and flying controls are normally operated by hydraulic mechanisms known as power control units (PCU), which are powered by the aircraft's hydraulic systems. Movement of the pilot's controls operates control valves (via cables) that determine the hydraulic input to the PCU and thus the degree of movement of the control surface. Since no direct connection exists between the control surfaces and the pilot's controls the force exerted by the airflow on a deflected control surface cannot be felt as a pressure on the control column or rudder pedals. The pilot thus has no direct feeling of flying control pressure when moving the pilot's controls and is at risk of stressing the aircraft by excessive demands.

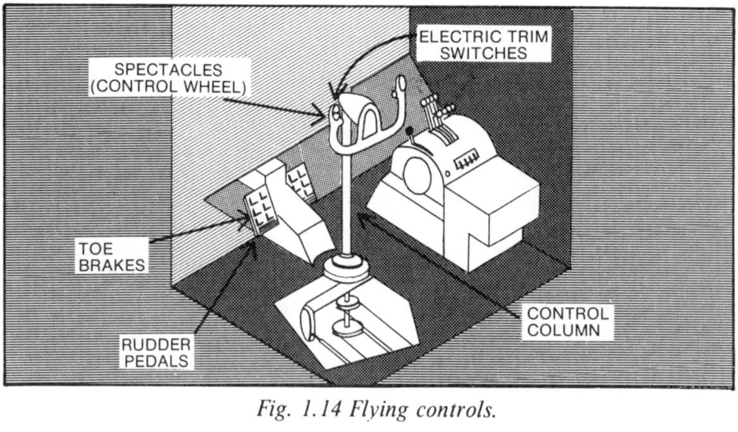

Fig. 1.14 Flying controls.

To overcome the problem artificial feel is supplied by 'feel' units, which apply pressure to the controls proportional to control surface movement. As a result the pilot obtains the sensation of flying the aircraft as if the controls were directly connected by cables to the control surfaces. Indeed, sometimes a degree of physical effort is required to overcome the realistic feel-unit pressures transmitted to the pilot's controls. With an outboard engine failure on take-off, for example, the swing can be quite marked due to the high engine thrust on one side, and directional control is maintained by rudder. It may take a few minutes for the drills to be completed before the pilot can settle down and apply the rudder trim to relieve the rudder control pressure, by which time the leg applying pressure to the rudder can be shaking with the strain.

Figure 1.15 shows the complete aircraft and the control surfaces.

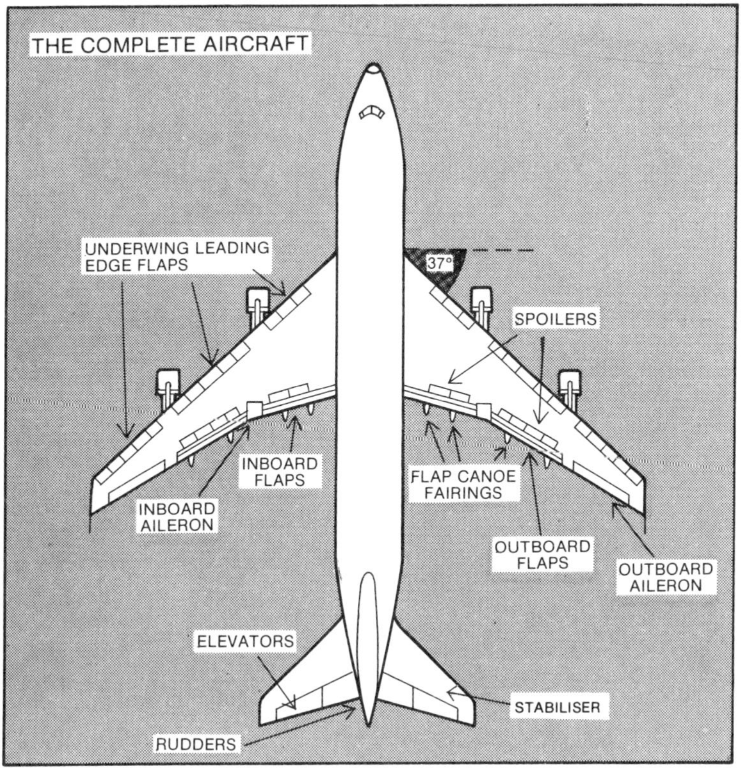

Fig. 1.15 The aircraft control surfaces.

Chapter 2
The Jet Engine

The earliest jet engine test bed run was conducted by Sir Frank Whittle in April 1937, but the first jet aircraft to fly was the Heinkel He 178 in August 1939. The Comet, the first jet transport, made its maiden flight in 1949, but regular transatlantic jet services did not commence until 1958 with the Comet 4 and the Boeing 707. The Boeing 747 made its first commercial New York to London flight in 1970, and Concorde made its first in the opposite direction, simultaneously from London to Washington and Paris to Rio de Janeiro, in 1976.

Principles

Although it is a complicated piece of machinery, the basic workings of the jet engine are, in fact, quite simple (Fig. 2.1). Air is drawn in at an intake by a compressor which highly compresses the air. The highly compressed air passes to a combustion chamber where it combines with burning fuel, expanding enormously. The fuel used is kerosene, which does not ignite instantaneously but which burns continuously like heating oil or paraffin. The expanding air from the

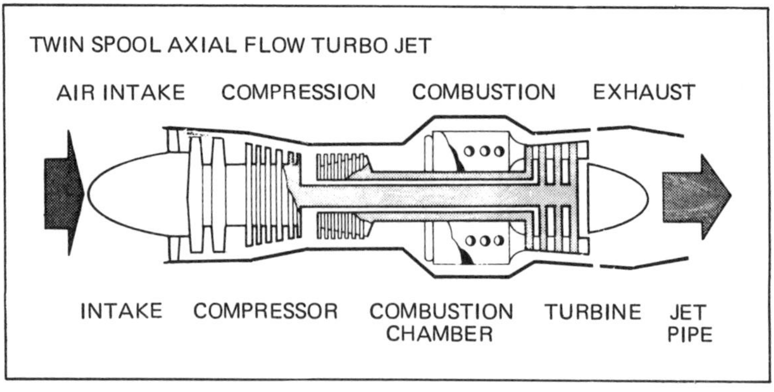

Fig. 2.1 The jet engine.

combustion chamber first channels through a turbine, which drives the compressor via a connecting shaft, before exhausting at great speed through the jet pipe. The flow of air through the engine, even at idle power, is such that a man can be sucked into the compressor within 8 metres (25 feet) of the intake, and can be be blown by jet blast within 45 metres (150 feet) of the jet pipe.

The jet engine cycle

Jet engine operation is a continuous cycle. Air is first drawn by the compressor into the engine intake. One stage of a compressor consists of a ring of rotating blades (known as rotors) followed by a ring of stationary blades (known as stators). The rotating rotor blades propel the air through the stationary stator blades with a resultant increase in pressure. The pressure increase across each stage is relatively small so that a number of stages are necessary to produce the required pressure. On larger jet engines airflow through the compressor is improved by breaking the compressor down into two or three separate sections known as spools; each spool being driven independently by its own turbine and connecting shaft. Compressors are denoted by the letter 'N', and compressor spools as N1, N2, (and N3). N1, therefore, corresponds to the low pressure (LP) compressor spool at the intake and N2 (or N3) to the high pressure (HP) compressor spool before the combustion chamber.

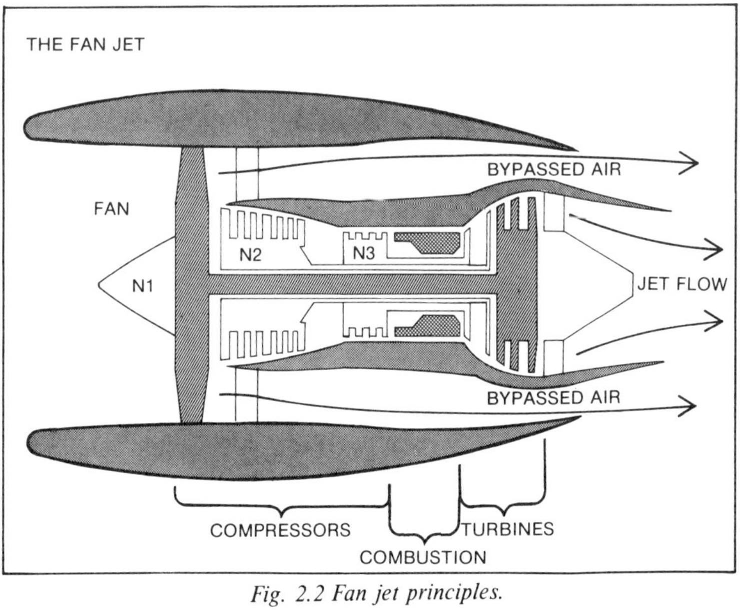

Fig. 2.2 Fan jet principles.

An improvement to propulsive efficiency is also achieved on large engines by arranging for some N1 compressor air to bypass the main engine core and to exhaust straight to the atmosphere via a bypass duct. Such engines are known as bypass engines (Fig. 2.2). Today's big jet engines have developed the bypass concept to such a degree that modern N1 compressors now consist mainly of a giant single ring of large blades, known as a fan, similar to a large many-bladed propeller with the tips cut off (Fig. 2.3). And, indeed, the fan is more like a propeller than a compressor, delivering 75 per cent of the thrust of the complete engine and bypassing five parts of air for every one that flows through the main engine. Engine development has now turned full circle and returned to the principle of the propeller!

Fig. 2.3 Front view of the fan jet.

The portion of fan air that enters the main engine core progresses through the N2 (and N3) compressors and discharges as hot, highly compressed air into combustion chambers where about one third combines with burning fuel, combusting at a temperature of around 2000°C, while the remainder is used for cooling. The expanding exhaust flow from the combustion chambers channels through stationary convergent guide vanes which direct the flow onto the turbine blades. The turbine rotates under the force of the airflow impinging on the turbine blades, and in turn rotates its respective compressor via the connecting shaft. Aft of the turbine the air

Fan exhaust shrouds the main jet core.

Fan jet with covers removed.

continues to expand as it flows through the convergent duct of the jet pipe and exhausts from the engine as a high speed jet.

At take-off, the enormous static thrust produced by a fan jet of around 50,000 lbs. (depending on type) results in a radial force on *each* blade equivalent to six fully laden London buses, and the power produced by an *individual* high pressure turbine blade (about the size of a credit card) equivalent to a Formula One racing engine.

Most jet engine noise is due to the shear effect of the high speed jet of air cutting the atmosphere. An added advantage of the fan jet is the reduction in engine noise due to the bypassed air shrouding the main jet core and lessening the shear effect. Aircraft noise is measured in perceived noise decibels (PNdB), which is a measure of the type as well as the level of noise. The Boeing 747 scores about 107 PNdB at heavy-weight take-offs and landings. Airport noise limits are normally in the region of 110 PNdB maximum during the day and 102 PNdB maximum at night. Engine noise-abatement techniques involve reducing take-off power to climb power at a certain height, usually 1500 feet above the departure airport.

Fig. 2.4 Thrust levers.

Engine start

The engine is first turned at speed by a small pneumatic starter motor to induce sufficient flow through the compressor. Fuel is then sprayed under pressure into the combustion chambers and igniters within the chambers are switched on to supply the initial source of ignition for the fuel. Once alight and burning, engine revolutions per minute (rpm) continue to rise until a point is reached at which the engine becomes self-sustaining. The starter is then disengaged and the igniters switched off. Engine acceleration continues until idle rpm is achieved. To accelerate beyond idle power the thrust lever is advanced on the flight deck (Fig. 2.4). The engine is shut down by simply cutting off the fuel supply.

In flight, restarting a shut down engine is achieved by maintaining an airspeed sufficient to create enough airflow through the engine to turn the compressor. The pneumatic starter can also be used, if required, e.g. at high altitude.

Engine instruments.

Engine performance

The performance of a jet engine is normally expressed in pounds (lbs) of thrust (*see* Principles of Flight p. 13). The propelling force of a jet is not the result of the action of the jet on the atmosphere but is an example of Newton's Third Law which states that 'for every action there is an equal and opposite reaction'. The action of the jet force

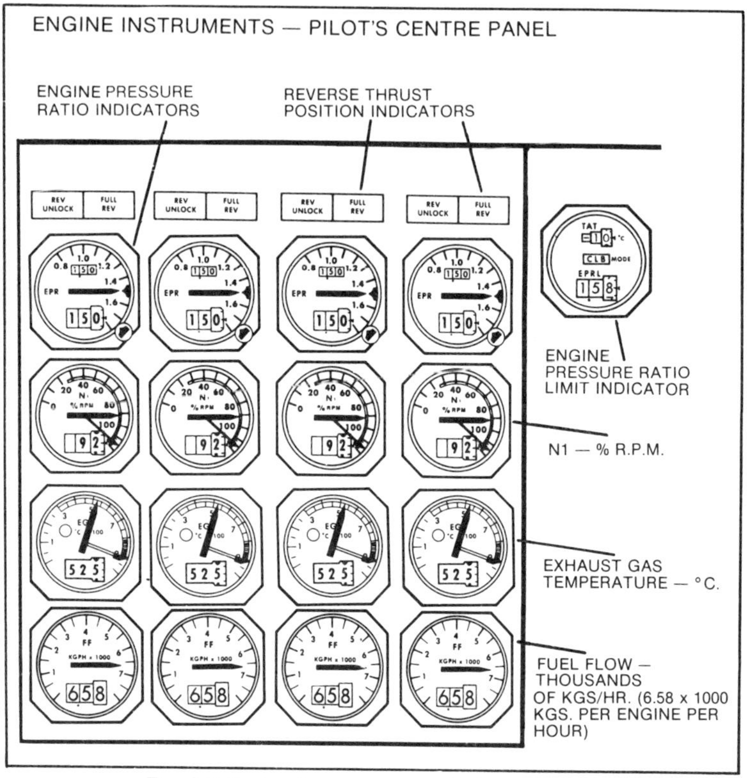

Fig. 2.5 Engine instruments – pilot's centre panel.

rearwards therefore results in a reaction within the engine that propels the aircraft forwards.

The continuous cycle of the jet engine results in an increased power production over that of the piston engine for a given engine size, and without the jet the large aircraft of today would not be flying. It has been estimated that the number of the highest powered piston engines of comparable size required for take-off for a Boeing 747 would be as many as 18, and to maintain normal cruise at high altitude would require considerably more.

On today's big jets power is indicated on gauges on the flight deck in terms of engine pressure ratio (EPR), which is the ratio of the turbine discharge pressure to the compressor inlet pressure. On the Boeing 747, for example, the Pratt and Whitney (PW-JT9D-7) engine in normal conditions has a full power take-off EPR indication of 1.44, and the Rolls-Royce (RR-RB211-524) 1.63.

Compressor speeds can be as high as 20,000 rpm, so for convenience are expressed as a percentage of the maximum; e.g. in cruise the normal N1 speed is approximately 90 per cent of the maximum rpm and is displayed as such on the gauge. The normal maximum operating speed of 100 per cent can be exceeded for short periods, and indicated rpm in excess of 100 per cent can actually be achieved on the gauges (e.g. maximum take-off N1 103 per cent).

The maximum EPR for take-off, climb, cruise and go-around is calculated by computer and displayed on an EPR limit gauge. Engine power must not be set beyond the maximum computed value as the engines are not mechanically limited and it is possible to overboost the engines and cause damage. On take-off and climb engine EPR is normally set at something less than maximum if aircraft weight allows, thus reducing engine wear and tear. Reduced power used in such cases is known as graduated power.

On descent the thrust levels are closed giving idle power, and the aircraft becomes, literally, a giant glider. On the approach and landing phases of the flight the thrust levers are handled more coarsely and engine power ratio (EPR) is set according to the speed required for each flight condition (e.g. approximately 1.05 - 1.10 on the final approach with gear and flap lowered). On the final approach to land, if the aircraft is required to climb away (known as a 'go-around') because of, say another aircraft blocking the runway, full go-around climb requirement is always set.

The engines are effective for take-offs from airports up to an elevation of 10,000 feet, and have a maximum permitted operating altitude of about 42,000 feet. Jet engine performance is proportional to the density of the intake air and therefore diminishes with increasing altitude as the air becomes thinner. Aircraft drag, however, also diminishes with altitude and at height there is actually an increase in speed despite lower engine performance. Also, jet engines operate more effectively at high rpm and only at high altitude can high rpm be set without producing excessive thrust. So jet aircraft fly fast and high (normally up to 41,000 feet), the loss of performance in the operation of the jet engine being more than overcome by the reduction in aircraft drag and better engine efficiency, with a resultant improvement in fuel consumption.

Engine failure

Engine failures are comparatively rare, although of course they do happen. The failure can be the result of any number of problems from turbine blade failure, to bird ingestion. Obviously, engine failure on take-off is the most likely owing to the high power settings involved, but even then the odds against failure have been calculated at 300,000 to 1. (The possibility of landing with a double engine failure has been calculated at one in a million flights.) If the engine failure occurs before V1 (the go or no go speed) the aircraft is brought to a halt by applying the brakes, deploying the speed brakes and selecting reverse thrust. The abandoned take-off procedure at speeds close to V1 is a dramatic event, and is only undertaken with such major incidents as engine failures. If the failure occurs after V1, there is normally insufficient runway length available for stopping and the aircraft is committed to take-off. A single engine failure on take-off after V1 can be successfully handled by the crew, the normal procedure being to select the gear up when the aircraft is safely climbing away and to carry out the engine fire drill. Climb is continued to 800-1000 feet above the airport, the aircraft levelled off and speed increased while bringing in the flap. The aircraft is now in a safe flying condition in the event of a second engine failure.

A single engine failure in the cruise can be dealt with successfully, although it may involve descent of a few thousand feet (perhaps more, depending on weight) into the more dense atmosphere to maintain cruise. Double engine failure presents a greater difficulty, but even then cruise can be maintained at the lower levels, say from 10,000 to 15,000 feet, depending on weight. (Of course, this could be a problem if the route lay across the Alps in Europe or the Rockies in

Fig. 2.6 Position of the 'fifth pod'.

the USA.) Approach and landing on three out of four engines is a well practised manoeuvre (on the simulator), and approach and landing on two engines, although more difficult, can be carried out with safety. On one engine, however, level flight cannot be maintained. At normal descent speeds, with engines at idle, the rate of descent is approximately 2500 feet per minute, so with one engine the descent rate would be somewhat less. With an instantaneous loss of all four engines at normal cruise height (a most unlikely event) the aircraft could probably remain airborne for about twenty minutes or so by reducing the speed to minimum.

For aircraft grounded by engine failure at a distant airport it is possible for a normal commercial flight to transport a spare engine out by slinging the engine in a pod below the left wing between the fuselage and inboard engine. A spare engine carried in this manner is known as a 'fifth pod' (Fig. 2.6).

Engine position

Over the years engine position has varied with jet design. On today's big jets the engines are either slung in pods below the wing, mounted at the tail, or a combination of both.

Wing pod engines

On aircraft such as the DC8 and Boeing 707 the engines are slung in pods below the wings, thus maintaining a relatively clean wing form. Since, in flight, the bending effect of lift on wings is upwards, engine weight acting downwards is a bonus, and the wings can be built with less strength and are therefore lighter. Disadvantages of wing mounted engines are the restriction to roll movement near the ground with the possibility of scraping a pod, especially on landing, and the problem of asymmetric thrust with engine failure, especially on take-off. If, after V1, while still accelerating along the runway, one of the outboard engines fails (i.e. number 1 or 4 — engines numbered from 1 to 4 from pilot's left to right), a marked swing is experienced which has to be corrected by rudder. Cross wind on take-off can also compound the problem. When the wind is blowing across the runway the tail fin acts as a vane, the wheels as a pivot, and the aircraft nose tends to 'weather-cock' into wind. If, on take-off, No. 4 engine fails with the wind blowing from right to left, the swing to the right could be marked and swift action would be required on behalf of the pilot to maintain the aircraft on the runway centre line (Fig. 2.7). With such a cross wind No. 4 engine is known as the 'critical' engine.

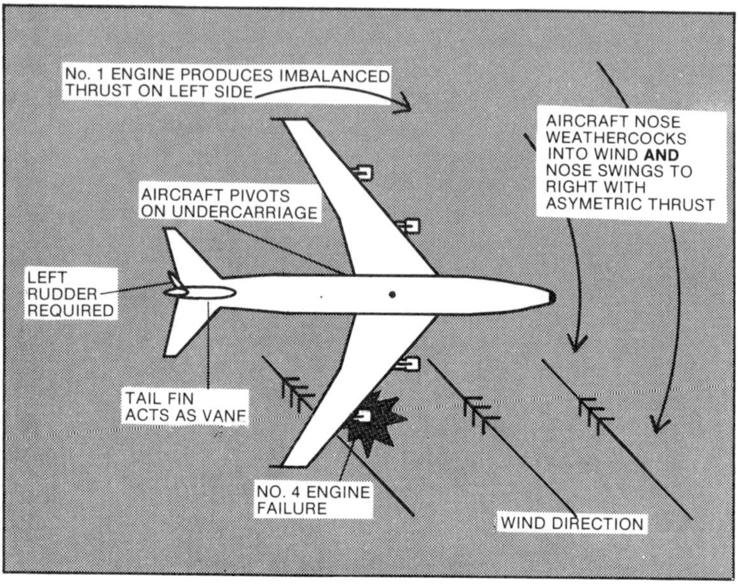

Fig. 2.7 Critical engine.

Tail-mounted engines

On aircraft such as the Boeing 727 and VC10 the engines are grouped together at the tail. This has the advantage of minimizing problems due to asymmetric thrust with engine failure but increases the danger, with severe engine failure, of another engine being damaged because of their close proximity. The wing form is very clean, but wing strength has to be increased because of lack of downward acting weight of the engines. Since engine noise is at the rear, the cabin is quiet, but with most fuel contained in the wings, fuel pipes feeding the engines have to run through the fuselage and are a potential hazard in the event of an accident.

Aircraft with engines mounted aft have the tailplane placed high on the fin, free of the engines, and have the potential to deep stall (for stall *see* Principles of Flight p. 16). Stall recovery requires speed, which is achieved by pushing the nose down (the effect of elevator on the tailplane) and by applying power. Wing form characteristics are such that at the stall the nose tends to pitch up. If the stall is allowed to develop, the engines drop into the turbulent flow from the wings. The disturbed airflow causes engine 'hiccup', known as a surge, with the risk of combustion ceasing (i.e. flaming out) and a resultant loss of power. If the stall develops still further the high tailplane drops into the turbulent flow and elevator control is lost. The aircraft is now said to be deep stalled (Fig. 2.8) from which recovery is difficult, if not impossible. Of course, such aircraft are well protected with 'stick shakers' and 'stick pushers', which shake and then push the control column forward at the onset of stall.

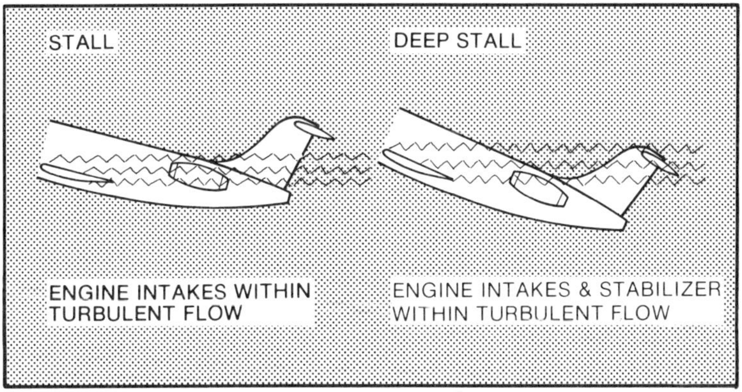

Fig. 2.8 Deep stall.

One final, rather obvious advantage of aircraft with rear mounted engines is that it is possible to install an odd number of engines (i.e. three), and it is not surprising that such modern big jets as the Tristar L1011 and DC10 have adopted a compromise between the layouts, with one engine being slung below each wing and one positioned in the tail.

Reverse thrust

Reverse thrust is used on landing and in the event of an abandoned take-off, as an aid to slowing the aircraft. (It is also used by some aircraft types in flight to reduce airspeed.) It is not, as some imagine, the engine somehow being reversed in direction, but on the big jets is a system of deflector doors that close in the path of the fan exit deflecting the airflow forward about 45 degrees (Fig. 2.9). On other jets, buckets clam together in the path of the jet efflux deflecting the jet flow forward.

The deflector doors or buckets are positioned within the engine fan exhaust or jet pipe, and in reverse deflect the airflow forward through grills or openings in the engine casing. Reverse thrust is selected by levers connected to the main thrust levers. With the main thrust levers closed, rearward movement of the reverse thrust levers first positions the blocker doors or buckets for reverse thrust operation, and further movement rearwards accelerates the engine to the reverse thrust power setting.

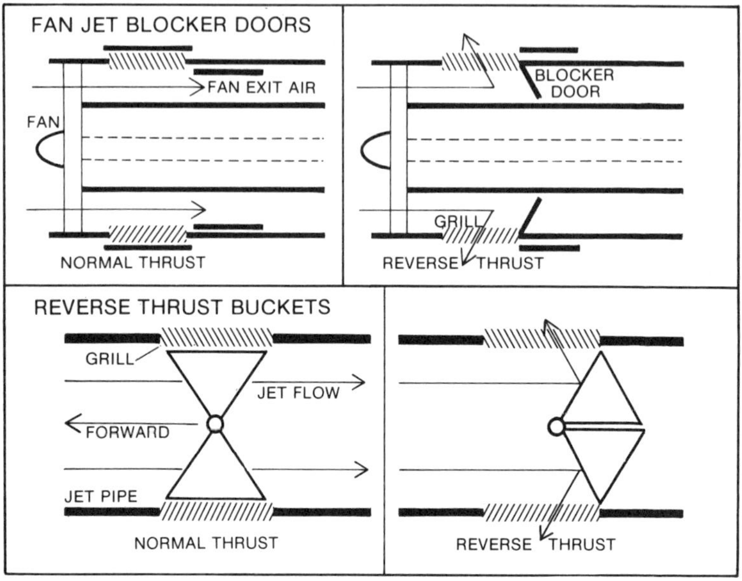

Fig. 2.9 Reverse thrust.

Exposed Reverse Thrust Grilles on the engine casings.

Auxiliary power unit (APU)

The APU is a small jet engine situated in the aircraft tail that can be started on battery power. The APU is run on the ground to supply electrical and pneumatic power to the necessary systems before engine start. During transit stops the APU is started before the main engines are shut down, and the aircraft can be powered independently of external power sources, (e.g. electrical power and air conditioning). The APU pneumatic source is normally used to restart the main engines.

Fuel

At a refinery crude oil is fed to a giant still where it is boiled up like a stew. The various petroleum derivatives separate out with the lighter constituents (gasoline and kerosene) being distilled off at the top, while the heavier constituents (diesel oil, heating oil etc., right down to the heavier industrial oils) remain at the bottom. This residue is then reheated to a higher temperature and further separation occurs. Continued processing and refining takes place until, eventually, each product, according to its properties, emerges ready for consumption. Gasoline (petrol), for example, with its property of rapid ignition, is used in piston engines, while kerosene (paraffin), being a burning fuel, is used in storm lamps and portable heaters, and in a more refined form in jet engines.

Auxiliary Power Unit situated in the aircraft tail.

The maximum fuel load of the Boeing 747 is approximately 39,000 gallons or 47,000 US gallons, or 177,000 litres, or 140 tons. The minimum allowable fuel load in service depends on company policy, but is around 22 tons. All fuel, where possible, is carried in the wings, the centre tank being used only when fuel requirements are above about 100 tons. Fuel in wings is a useful downward acting force (similar to the wing mounted engine pods) and the downward bending movement of the fuel weight is prolonged by first using centre tank fuel (if any), followed by inboard main then outboard

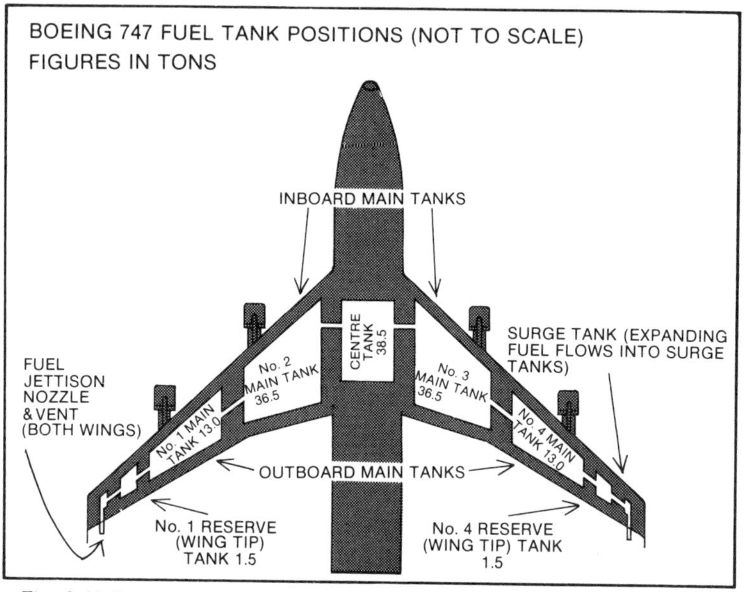

Fig. 2.10 Boeing 747 fuel tank positions (not to scale). Figures are in tons.

main wing tanks, and lastly wing tip tanks; all fuel switching and monitoring being accomplished by the Flight Engineer (Fig. 2.10).

Refuelling is normally from underground tanks situated below parking bays where fuel is pumped aboard via pumping vehicles which position by the aircraft. Occasionally, at remote stands, bowser tankers are required. Static lines are used to bond trucks and aircraft to the ground to prevent dangerous build ups of static during refuelling. The fuel line is connected to a point below the wing and fuel is pumped aboard, at the rate of 800 gallons per minute, to all tanks via internal interconnecting pipes. Since some fuel is normally remaining in the tanks, and only sufficient is loaded for the flight, average refuelling time is about 25 minutes. Gauges on a refuelling panel by the fuel point, and on the flight deck, indicate the quantity of fuel being loaded, and on completion are checked against the vehicle gauges to verify the load. A sample of fuel is also taken to check for evidence of water which, if above a certain limit, would freeze at altitude and block fuel filters.

The fuel required for each flight is calculated mostly by computer, but in some cases by hand, using the forecast wind and temperature, and expected flight levels. This information is displayed to the flight crew in the form of a fuel flight plan. For an average 7 or 8 hour flight (Europe to the United States, the Far East to Australia etc.) the

Fuel pumping vehicle.

Boeing 747 fuel requirement is approximately 80 - 90 tons. Diversion fuel, together with alternate reserve, is always carried (just in case) and is normally in the region of 10 to 15 tons, giving, for example, Boston as diversion for New York, Manchester for London, and Kuala Lumpur for Singapore. To this is added contingency fuel which, as the name suggests, may be used in the event of an upset to the fuel flight plan, such as an adverse change in the forecast wind, a lower flight level than expected being allocated (with resulting higher fuel consumption) or holding delays at destination caused by traffic or weather. Any fuel carried above the basic requirement is known as excess fuel. Fuel figures for a typical London - New York flight are shown overleaf:

	Fuel (tons)	*Time (hrs)*
Fuel and time to New York	80.0	7.00
Diversion fuel (Boston)	13.0	1.10
Contingency fuel	4.0	.25
Excess fuel	0	0
	97.0	8.35

The total fuel requirement from London to New York is therefore 97.0 tons, of which only 80.0 tons would normally be used, and the maximum time the aircraft can remain airborne with this fuel load, i.e. the endurance, is 8 hours 35 minutes.

It may be necessary, on occasions, to carry excess fuel if significant delays are expected or if severe or poor weather is forecast at the destination for the time of arrival. The decision to carry excess fuel, however, is not taken lightly, as any increase in weight (whether passengers, cargo or fuel) increases fuel consumption, and a significant portion (i.e. 3 per cent per hour) of any excess fuel loaded is actually used up in just carrying that excess. (It has been calculated that to carry regularly the weight of just one small sachet of sugar would increase fuel consumption by one gallon per year.)

Fuel Bowser tanker.

Flight Engineer's fuel panel.

On the other hand, of course, if delays or bad weather are expected it is essential to carry sufficient fuel, as civilian aircraft cannot be refuelled in mid air, and once airborne the aircraft is on its own. On 20 October, 1979, a Boeing 747 was diverted from its destination, New York, to Newark International in New Jersey, some 30 miles away, after holding overhead Kennedy for some time due to bad weather. The number of aircraft diverting involved the Boeing 747 in a 100 mile detour for traffic sequencing and, with fuel critically low, the Captain declared an emergency. Just after touchdown, with the aircraft decelerating down the runway, the fuel ran dry, and as engines were unreversed, Nos. 1 and 4 flamed out from fuel starvation. The aircraft cleared the runway, and after taxying about one mile No. 2 engine ran down. In fact, this incident was the result of the use of an erroneous fuel chart, but it illustrates well the problem of aircraft fuel running low. Fortunately, such incidents are rare.

The two most useless things in aviation are said to be 'runway behind the aircraft' and 'fuel left in the bowser', but the oft-heard pilot adage of enough fuel for the flight and a 'bit for mum' has had to be modified in the light of fuel shortages and soaring prices. Pressure is on Captains to carry the lowest fuel requirements commensurate with safety and the law, but with unpredictable bad weather in certain areas, and frequent unexpected delays due to increased traffic in others, the dividing line between sufficient fuel and running out in flight can be very thin indeed. Not surprisingly, crews tend to err on the side of safety. As they say, fuel on the ground is expensive, but fuel in the air is priceless.

On occasions, the opposite is the case, and the aircraft is unable to carry sufficient fuel for the planned flight. It may be that a reduction in maximum take-off weight is required (as at hot and high airports such as Nairobi) or excessively strong headwinds are forecast *en route* (as sometimes experienced on long distance flights from Europe to the US West Coast). In such cases a technical stop *en route* is normally required for refuelling, but occasionally it is possible to flight-plan again in the air to enable the destination to be reached non-stop. A typical route is Nairobi-London where it may not be possible to load sufficient fuel for the trip. In this case Rome is nominated as the refuelling stop with, say, Frankfurt as diversion, and any extra fuel able to be loaded is shown as excess. Fuel savings *en route* are not unusual and, approaching Rome, may be sufficient to continue the flight by nominating Frankfurt as destination with Paris as diversion. Approaching Frankfurt, especially in good weather and with so many airports available in Europe, it is now possible (and legal) to reduce reserves and to nominate Paris as destination with London as diversion. Once again, approaching Paris, with good weather in the London area, it may be possible further to reduce the fuel required by nominating a closer than normal alternate to London, say London Gatwick. Sufficient fuel can then be found to nominate London as destination and Gatwick as diversion, and thereby reach London direct. The procedure is, of course, purely a paper exercise, but to remain legal at each recalculation point sufficient fuel must be shown to be available for the remainder of the flight. As can be imagined crew workload, especially during the latter part of the flight, is very high due to the constant re-planning required.

At the other end of the scale, the dumping of fuel in flight is possible and is a simple process. In an emergency, especially with an engine failure shortly after take-off necessitating a return to the departure airport, the aircraft weight can be reduced down to landing weight by dumping fuel. Valves can be opened to allow fuel to be jettisoned at the rate of two tons per minute by pumping via nozzles on the trailing edge of each wing tip, allowing a significant weight decrease to be achieved rapidly. (See photo p. 59).

In normal cruise conditions, the required engine power setting and aircraft speed are extracted from graphs against height and all-up weight. At the start of the flight the gross weight indicator is set to the aircraft take-off weight and runs down as fuel is used, giving a continuous readout of aircraft weight. As weight decreases with fuel consumption, engine power setting is reduced accordingly until a point is reached at which the aircraft is light enough to climb to the next higher *en route* flight level (normally a 4000 feet jump) resulting in improved fuel economy. Fuel consumption in the cruise varies with weight and altitude, but for the Boeing 747 is approximately 10-12 tons per hour (about 3000 gallons, or 3600 US gallons, or 13,700 litres per hour, or just under 6 gallons per mile, or one gallon every second). With full power take-off at maximum weight fuel flow per engine approaches 8 tons per hour giving a total fuel consumption in excess of 30 tons per hour. Full power, however, is only applied for about two minutes as engines are throttled back to climb power at 1500 feet, and the fuel flow per engine drops to about 6.0 - 6.5 tons per hour. (Taxying uses about one ton of fuel, so taxying, take-off and climb to 1500 feet consume approximately two tons.) Fuel consumption continues to reduce with climb, dropping to a total of about 16 tons per hour at 30,000 feet, and to 12 tons per hour when maintaining cruise at 31,000 feet.

In cruise the Flight Engineer makes frequent checks on everything from fuel tank temperatures to fuel flows. At height the outside air temperature (OAT) can be as low as -60°C or -70°C and fuel heat is required to prevent engine fuel filters from icing up. The fuel itself has a freezing point of -40°C to -50°C, and in extreme cases it may be necessary to descend to a lower and warmer altitude if tank temperatures drop too close to freezing point. At intervals checks are also made on the fuel state. The Flight Engineer sums fuel gauges to

Got the final fuel figure, Guv'nor?

(SEO Chris Baxter, Tristars,B.A.)

obtain the fuel on board and calculates from tables the fuel required to destination. These figures are then compared against fuel flight plan requirements. Any unexpected problems, such as strong head winds or lower than requested flight levels, which may reduce reserves, become readily apparent during these fuel checks, and early action can be taken.

Fuel conservation is now of major interest to the world's airlines as fuel costs are almost one third of total operating expenditure, and, cost apart, fuel is rapidly running out. Estimates state that oil reserves will be used up in 30 years' time. Aviation consumes only about 4 per cent of the world's total oil consumption, but at present no other fuel is available for the industry to survive. Liquid hydrogen is the best possible alternative, and plans are laid already to test the fuel operationally on cargo aircraft, but its widespread use is many years away.

Chapter 3
Radio and Radar

Basic radio theory

An alternating current (a.c.) is one in which the direction of flow is being constantly reversed at regular intervals. When a graph is drawn of current against time a sine wave pattern is produced (Fig. 3.1). The current is seen to start at zero and increase to a maximum in one direction, then pass through zero to a maximum in the opposite direction, and once again back to zero. This sequence is known as a cycle, and the peak values as the amplitude of the current. The number of cycles occuring in one second is known as the frequency and is expressed in cycles per second, or more recently in Hertz (Hz) in honour of the German physicist of that name who lived in the latter part of the nineteenth century.

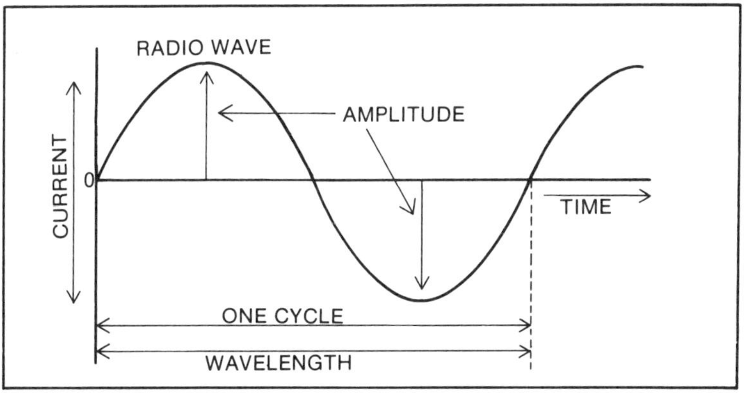

Fig. 3.1 Sine wave.

If an a.c. current is fed at the required frequency to a suitable antenna, the energy is not contained within the antenna but is radiated out into space in an electromagnetic form known as a radio wave. This energy transmitted through space comprises both an alternating electrical field and an alternating magnetic field, positioned at right angles to one another. A vertical antenna, for example, produces a mainly vertical electrical field with a horizontal magnetic field. Such a signal is known as a vertically polarized radio wave (Fig. 3.2). To obtain efficient reception the receiving antenna is also required to be vertical.

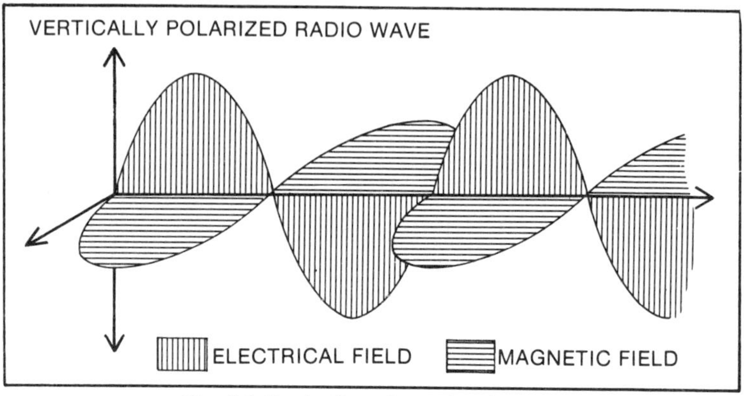

Fig. 3.2 Vertically polarized radio wave.

Frequency

A frequency of one cycle per second equals one Hertz; 1000 Hertz equals one kiloHertz (kHz); 1000 kiloHertz equals 1 megaHertz (MHz). Aviation transmissions are expressed in terms of kiloHertz or megaHertz.

Wavelength

Wavelength is the distance travelled by a radio wave during the transmission of one cycle, and is expressed in either metres or centimetres.

Frequency and wavelength

Frequency and wavelength are related; the connection between the two being the speed at which radio waves travel — the speed of light.

Frequency (Hz) × wavelength (metres) = Speed of light (metres/second).

E.g. to calculate the wavelength of a 200 kHz transmission:

$$\text{Wavelength (metres)} = \frac{\text{Speed of light (metres/second)}}{\text{Frequency (Hz)}}$$

$$= \frac{300{,}000{,}000}{200 \times 1000}$$

$$= 1500 \text{ metres}$$

Frequency bands

Frequency bands encompass a specific frequency range. Each band has its own properties of transmission suitable for such uses as communications, navigation aids or radio beacons, etc. Aviation communication frequencies, for example, lie within the high frequency and very high frequency bands.

Very low frequency (VLF):	3 - 30 kHz
Low frequency (LF):	30 - 300 kHz
Medium frequency (MF):	300 - 3000 kHz
High frequency (HF):	3000 - 30,000 kHz
Very high frequency (VHF):	30 - 300 MHz
Ultra high frequency (UHF:	300 - 3000 MHz
Super high frequency (SHF):	3000 - 30,000 MHz

Phase

The phase of a current is the stage reached in a cycle at a given instant and is expressed in degrees from 0° to 360° (Fig. 3.3). Two radio waves transmitted on the same frequency can be in or out of phase with each other. Any phase difference between the two can either be used to advantage, as in certain navigation equipment, or can cause interference, as in VHF communications.

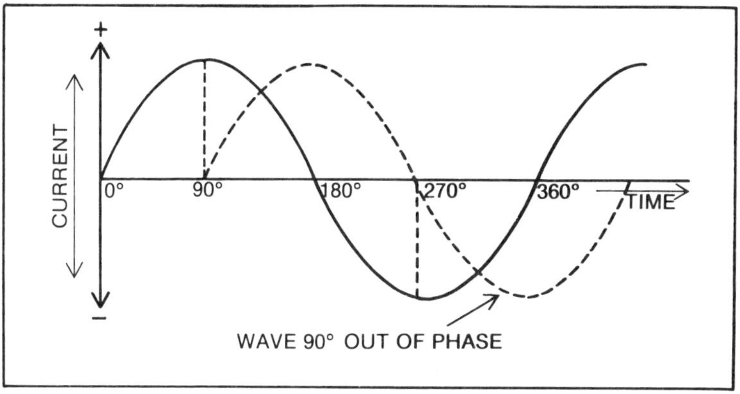

Fig. 3.3 Phase relationships between two waves.

Propagation paths

Signals transmitted from an antenna scatter radio energy in all directions. The paths in which the radio waves travel to a receiver follow a variety of different routes depending on the power of the transmitter, the distance of the receiver from the transmitter, and the frequency of the transmission. Radio signal propagation paths can be divided into two basic wave forms — the ground wave and the sky wave.

Ground wave

The ground wave can be further subdivided into the direct wave, the ground reflected wave and the surface wave. Ground waves travelling in a direct line from transmitter to receiver are known as direct waves, and those reflected from the ground before being received, as ground reflected waves. The direct wave transmission is known as 'line of sight' in that the receiver is 'seen' in a direct line from the transmitter (i.e. some radio waves can pass through most buildings or structures but not through mountains or over the horizon). Since the direct and reflected waves follow differing paths, they may be received out of phase causing fading or temporary loss of signal. Under normal conditions, only 'line of sight' reception can occur above a frequency of 30 MHz in the VHF and higher bands.

Ground waves that closely hug the surface of the earth are known as surface waves. They are a feature of the lower frequency bands where a phenomenon known as diffraction occurs, producing strong radio waves that can be transmitted over great distances. Surface waves also appear in the higher frequencies but the range is limited to only a few miles.

Sky wave

Sky waves are radio waves reflected from layers within the ionosphere and bounce back to earth over great distances. The sun's ultraviolet light causes electrons to become separated from gaseous molecules in the atmosphere resulting in positively charged molecules known as ions. In the ionosphere, stretching from about 50 km to over 400 km above the earth's surface, distinct layers of ions form known as 'ionized layers'. During the day four main ionized layers are present, but at night, when the sun's ultraviolet radiation is absent, only two distinct layers remain. The existence in daylight of the lower ionized layers results in radio waves being absorbed, producing weak sky wave reflections. At night the absorbing layers disperse and copious sky wave reflections occur in the HF, and lower frequency bands, providing increased range and improved reception.

Where a radio wave is reflected from an ionized layer and bounces back from the earth's surface to be reflected once again, and so on, ranges of 8000 nautical miles or more can be achieved. In the VHF band, and at higher frequencies, the wavelength is too short to allow reflection from the ionized layers and sky waves seldom occur.

Propagation path summary
Figure 3.4 illustrates four situations:

1. The aircraft is receiving both direct and (perhaps) ground reflected waves.
2. The aircraft is receiving the surface wave only.
3. The aircraft is receiving the sky wave only.
4. The aircraft is not receiving any signals.

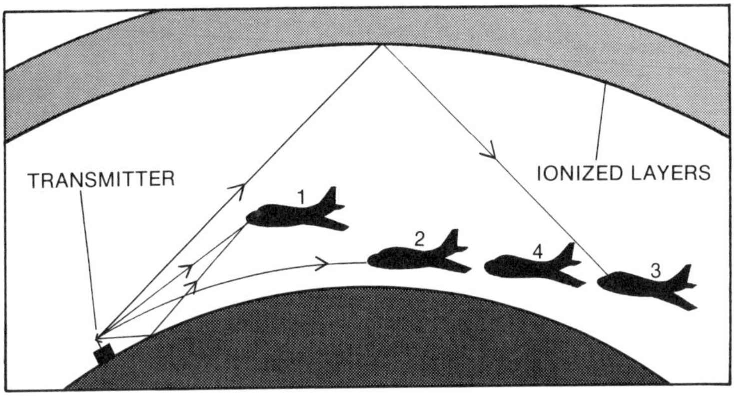

Fig. 3.4 Propagation paths.

In situation (4) transmissions cannot be received as the aircraft is over the horizon (direct line of sight signals cannot be received), is outside the range of the surface wave, and is inside the range of the first sky wave. Selection of another frequency may improve reception. If not, the pilots simply have to wait until the aircraft is positioned within the reception area of the frequency in use.

Frequency band properties
Each frequency band has its own properties that make it suitable for particular uses. In the VLF and LF bands the strong surface waves and sky waves produced provide long range signals for navigational equipment. Most broadcast stations are found in the MF band where the range is more suitable and static interference acceptable. In the

HF band the extreme range available is very useful for long range communications, but static is apparent to a greater or lesser degree. VHF is limited in range but provides clear, almost static free, short range communications. In the UHF and SHF bands the short wavelengths available are suitable for the pulse type transmissions used in radar and certain navigational equipment.

All aviation radio transmissions fall within the HF and VHF bands. In the VHF band commercial and domestic station frequencies range from 88 to 108 MHz, and aviation frequencies from 108 to 135.95 MHz. Aviation enthusiasts therefore require a special radio incorporating the aviation VHF range in order to listen to aviation transmissions.

Types of radio emission

Radio transmitters produce different types of emission depending on the signal required to be transmitted; the simplest emission being a continuous wave of constant amplitude transmitted on a fixed frequency. This simple radio signal carries no intelligence in the form of morse, music, speech or other information. In the communication frequency bands the radio frequencies used are too high to produce an audible tone, the human ear being only able to hear frequencies of up to about 8 kHz. The audio frequencies are too low for efficient transmission, so the difficulty is overcome by combining a radio frequency with an audio frequency at the transmitter and transmitting the resultant wave form through space. The receiver then separates the two frequencies and feeds the audio frequency to a speaker or headset.

This combining of frequencies is know as modulation, and the radio signal carrying the audio frequency as the carrier wave. The two types of modulation used are amplitude modulation (AM) where the frequency of the carrier wave remains constant and its amplitude is made to vary, and frequency modulation (FM) where the amplitude of the carrier wave remains constant and its frequency is made to vary.

Civil aviation communications are amplitude modulated. In the VHF band transmissions are relatively static free, but radio 'noise' is a problem with HF signals. Combining of the radio and audio waves for transmission results in a small spreading of the carrier wave frequency known as the bandwidth. An improvement to AM signal quality on HF radio is obtained by suppressing one side of the bandwidth at the transmitter. The resultant narrowing of the bandwith cuts down on the amount of radio noise and increases the effective range. Such transmissions are known as single sideband

(SSB), one side of the signal being known as the upper sideband (USB), the other as the lower sideband (LSB).

Frequency modulated transmissions are used for certain navigation equipment. Since radio noise effects amplitude more than frequency, FM signals are interference-free and are used in VHF radio broadcasting.

VHF radiotelephony (R/T)

All jumbo flight decks are fitted with three VHF radios. Frequencies are dialled by selectors and presented as digital displays allowing precise tuning.

Communications within 200 n.m. of an air traffic control (ATC) station are conducted on VHF. Outside this range HF communications are required, unless relay stations are positioned along route, as in some remote areas where signals are retransmitted down the line to the ATC centre. Reports can also be relayed via other aircraft: messages are passed from one aircraft out of range to another within range of a station. Here the camaraderie of the air extends beyond international boundaries, and Aeroflot passes messages for Pan Am, Olympic for Turkish Airlines, and Pakistan International for Air India, etc.

Normally position reports are sent at each reporting point along route, unless the area is under radar control, as in the USA and Europe, where position reports are usually omitted. When transmitted, a position report consists of the actual time of arrival (ATA) at the reporting point, the flight level of the aircraft and the estimated time of arrival (ETA) at the next reporting point.

As a flight progresses the pilot is passed from one control to another and at times the frequency changing can be quite rapid. On a short flight within Europe, for example the one hour hop from Frankfurt to London, as many as a dozen or more frequency changes can be involved. As only one person, either pilot or controller, can talk on the radio at a time, each has to take his turn to speak, and frequently the volume of chatter is quite considerable. When two talk at once a distinct squeal is heard. (Most people on hearing a busy aviation frequency for the first time are amazed by the flow of conversation.) Each pilot listens out for his own callsign, usually followed by certain instructions, for example 'Clipper One, turn right heading three two zero', and then repeats the message before complying. (Examples of airline callsigns are: British Airways — Speedbird; Cathay Pacific — Cathay; Pan American — Clipper; Trans World Airlines — T.W.A. — Aer Lingus — Shamrock.) 'Heavy' is added to callsigns of aircraft of Boeing 707 size and above, e.g. Speedbird Five Heavy.

Some of the R/T phrases are strictly standardized: for example, 'over', signifying a message is completed and a response is expected; 'out' signifying the end of a communication (never 'over and out'), etc. However, in the cluttered and congested frequencies of today the standard R/T phraseology has become somewhat modified in practice, although certain specific standard terms are still in use, e.g. 'roger' - message received; 'wilco' - will comply with instructions; 'expedite' - hurry; 'affirmative' - yes; 'negative' - no. The lack of total standardization, however, does not detract from the quality of R/T, and, indeed, aviation R/T is of the highest professional standard. There is a certain quiet politeness mixed with the clipped and precise messages. 'Good morning, Frankfurt, this is Shamrock One, flight level three five zero'. 'Good morning, Shamrock One, radar identified, omit position reports'. In the USA it tends to be a bit less formal and much more lively, mainly due to the volume of traffic.

English is the international language of the air and must be available, but the use of English is not compulsory and in many European countries — France, Italy, Yugoslavia, Turkey — the local languages are widely used. Even if one can speak English, learning the form of R/T terms and phrases and adapting to the many speech peculiarities of countries is something like learning another language. Sometimes there is even difficulty in understanding the plain English used, especially in such countries as Japan where pronunciation is a problem.

Spelling on R/T is accomplished using what is known as the phonetic alphabet. The old phonetic alphabet consisted mainly of English boys' names of which only Roger remains, but is now used in pilot R/T vocabulary. The phonetic alphabet of today has a more international flavour, as can be seen from Fig. 3.5. All pilots can spell quickly and easily (although perhaps, not always correctly) using the phonetic alphabet. Most pilots try to use the local pronunciation of radio beacons or position reporting points where possible, but when difficulty arises the identification letters can be given instead using the phonetic alphabet. For example, the radio beacon Bozhurishte in Bulgaria, identification code B O Z, can simply be pronounced bravo oscar zulu. It is also a requirement for pilots to learn morse, as all radio beacon identification codes are transmitted in morse code.

As an aircraft progresses along route, one pilot maintains a flight log noting all arrival times at reporting points, flight levels flown, estimates for the next, and the radio frequency in use. Normally No. 1 VHF is used for communication, No. 3 for the VHF emergency frequency of 121.5 MHz, which is monitored at all times, and No. 2 for another suitable station. Over busy areas such as the Middle East both Nos. 1 and 2 VHF sets are used for communications. Over

	PHONETIC ALPHABET	MORSE			
			M	Mike	— —
			N	November	— •
A	Alfa	• —	O	Oscar	— — —
B	Bravo	— • • •	P	Papa	• — — •
C	Charlie	— • — •	Q	Quebec	— — • —
D	Delta	— • •	R	Romeo	• — •
E	Echo	•	S	Sierra	• • •
F	Foxtrot	• • — •	T	Tango	—
G	Golf	— — •	U	Uniform	• • —
H	Hotel	• • • •	V	Victor	• • • —
I	India	• •	W	Whiskey	• — —
J	Juliet	• — — —	X	X-ray	— • • —
K	Kilo	— • —	Y	Yankee	— • — —
L	Lima	• — • •	Z	Zulu	— — • •

Fig. 3.5 The phonetic and morse alphabets.

Africa No. 2 VHF set is selected to 126.9 MHz and used on routes from Europe to Southern Africa. In terms of communications Africa is still the 'dark continent' and radio communications are frequently bad. On occasions entire countries are crossed without the flight crew being able to contact a single controller. As a precaution pilots transmit position reports to each other on 126.9 MHz. If a crisis arises without controller contact flight crews simply arrange their own separation on this frequency.

One VHF set is also often tuned to VHF meteorological broadcasts. At most major airports local conditions of wind, temperature and altimeter pressure setting are broadcast by an automatic terminal information service (ATIS), e.g. Frankfurt 108.2 MHz. *En route* centres, especially over Europe, also broadcast weather on VHF for a number of destinations, examples being Brussels on 127.8 MHz, Geneva on 126.8 MHz and London on 128.6 MHz. Companies also have their own VHF frequencies for communication between aircraft and their representatives. Frequency 123.45 MHz, for simplicity, has been adopted as a general chit chat channel between pilots.

In the event of an emergency a distress message is transmitted on the ATC frequency in use at the time, or, if no contact, on the emergency frequency of 121.5 MHz and is preceded by the word 'Mayday' (from the French expression 'M'aidez' — help me) repeated three times. The aircraft in distress then passes such information as the callsign of the flight, the nature of the distress, the position and height, and, if possible, the intentions of the pilot.

Complete radio failure is a rare occurence with modern aircraft, but, in that event, procedures are available for the guidance of a flight. Countries have different ideas on procedures, but normally a flight is able to continue to its destination in radio silence. If the failure occurs just after take-off, in clear conditions, the pilot can return to the airport of departure. If in instrument conditions, the flight maintains the last cruising level until clear of that portion of the route to which the level had been assigned, and then simply climbs to the cruising level indicated on the flight plan. This height is then maintained until over the holding point of the destination airport. The crew should attempt to arrive over the holding point as close to the estimated arrival time as possible on the flight plan, and then commence descent within 10 minutes of arrival at the holding point, and land within 30 minutes.

Many funny incidents occur during R/T communications and every pilot has his own list of favourite stories. Here are just two.

Two KLM (Royal Dutch Airlines) aircraft were flying in opposite directions within the Bombay control area, and when passing each other a few words were exchanged in Dutch on the control frequency. Communications in the area are not good. Often it's difficult to contact Bombay and frequently messages have to be repeated. On hearing the Dutch (not exactly the softest of languages to listen to) the Bombay controller simply assumed that an aircraft was attempting contact. At the end of the conversation between the aircraft the Bombay controller transmitted, 'Aircraft calling Bombay, say again please, you're very garbled'.

In Manchester, the crew of a British Airways aircraft, waiting at the runway holding point, were watching the landing of an F27 (Fokker Friendship) before they could take off. Unfortunately the aircraft bounced very badly, and a voice from the BA flight was heard to transmit, 'Oops, that was a bit of a Fokker!' Almost immediately, came the quick reply from the F27, 'Yes, it was nearly the end of a beautiful friendship!'

HF R/T

HF radio is used for all long distance communications. Each transmitting station is allocated about six frequencies in an attempt to cover the required range and the differing conditions of day and night. Generally the higher frequencies are required by day, or at

H.F. Antenna (one on each wing tip). Also seen is the fuel jettison nozzle and static wicks for discharging static electricity in flight. (Similar on right wing).

greater range, and the lower at night, or closer range. The primary frequency allocated for communication depends on the circumstances, but a secondary frequency is usually also nominated for use in case of difficulty. Not infrequently, stations cannot be contacted on any frequency. Over the North Atlantic, the Pacific, Australia, New Zealand and the Far East, especially when using SSB, HF communications can be very good, but over much of Asia and Africa it's a different story. Outdated and badly maintained equipment on the ground is used in an attempt to maintain contact with modern aircraft. Often static is so severe that pilots literally have to shout the messages over the air, when and if they can make contact. The one suitable frequency might be used by a number of stations across the globe, and voices can be heard simultaneously calling Cairo, Khartoum, Bombay, Bahrain, Colombo, and even as far as Singapore. The passing of a simple position report can become quite impossible, with others' messages travelling to and fro for thousands of miles on the same frequency. Communications on HF R/T tend to be more formal than on VHF because of the frequent difficulties. Pilots tend to use such terms as 'go ahead' at the end of messages instead of the recommended 'over' and 'charlie' instead of 'affirmative' as they seem to transmit more clearly over the air.

Fortunately pilots don't have to sit with headsets on and listen continuously to this cacophony of sound. A system known as Selcal (pronounced sell call) is used, whereby pilots can be alerted to a station calling them. Each aircraft has its own four-letter code, like a telephone number, BD-KL, which a controller can select to transmit a signal and activate the system. A 'ding dong' sound is heard on the flight deck and amber lights flash. The pilot can then replace his headset and answer the call. There are occasions, of course, when Selcal does not operate, and the pilot has no choice but to listen out as normal.

One controller operating an HF frequency controls a vast area. On the Atlantic the Shannon and Prestwick areas are combined to form one control area known as Shanwick, which extends out to 30° West. The other side, from Greenland to the Caribbean is controlled by Gander, on the Canadian coast, and New York. Examples of the Gander and Shanwick frequencies are: 2945 kHz, 5638 kHz, 8854 kHz and 13288 kHz, upper sideband (USB).

HF is also used for transmitting airport weather forceasts; for example, on the Atlantic, once again from Shannon (8833 kHz, and 5533 kHz) for all European destinations, and from New York (5652 kHz and 8868 kHz) for all East Coast US destinations. Time signals, too, are transmitted throughout the world on HF. Most nations broadcast worldwide on HF, examples being the BBC World Service and The Voice of America, both of which are excellent sources of news for crews abroad.

Basic radar theory

Primary radar

The word radar derives from the term 'radio detection and ranging'. The principle involves a transmitted radio signal being reflected from an object to produce a weak echo, the strength of which depends on a number of factors, including the power of the transmitted signal, the shape and material of the reflecting surface, and the size and range of the 'target'.

Since the speed of propagation of the transmitted signal is known, the range of the target is simply determined by measuring the time difference between the transmission of the signal and reception of the echo. If a continuous radio signal were transmitted it would not be possible to measure this time difference, so radar transmitters emit a series of very short bursts of radio energy called pulses, and receive any reflected echoes from a target in the gaps between pulses, thus allowing measurement of the time interval from the start of the transmitted pulse to the start of the corresponding received echo pulse. The pulse length is of necessity short, to prevent the end of the transmitted pulse obscuring the start of the much weaker echo pulse, especially when receiving from a target at close range. Radar signals form a narrow beam, which is usually transmitted from a scanner moving in a horizontal or vertical motion, the echoes received being presented as a 'blip' on a radar screen indicating both the direction and the range of the reflecting target. Examples of primary radar are airborne weather radar, where the echoes are reflected from large water droplets evident in storm clouds, and precision approach radar

(PAR) which is used to monitor the approach of aircraft at certain airports in bad weather. The radio altimeter also works on the same principle.

Secondary surveillance radar (SSR)
SSR involves a ground transmitted radar signal being received on board an aircraft by a small receiver/transmitter known as a transponder, which responds by transmitting a second signal which is in turn received by the ground based radar receiver. The resultant signal received on the ground is much stronger than the weak echo detected using primary radar. An added advantage is that the transponder transmits on a frequency different from the ground based radar transmitter. Since the radar receiver is tuned to the transponder frequency, weak echoes of the transmitted radar signal that may be reflected from the target or from storm clouds in the vinicity are eliminated on the radar screen. A distinct clear image of the target is displayed. The transponder can also transmit coded signals that positively identify the aircraft on the radar screen. Almost all air traffic control in congested airspace today involves aircraft fitted with transponders, and many countries now require aircraft operating above a certain altitude to be equipped with transponders before entering controlled airspace.

Radar operation

Surveillance radar
This operates as described under primary radar. The aircraft has first to be positively identified, and this can be achieved by the pilot confirming his position from another source, such as a radio beacon. The bearing and distance from the beacon, stated by the pilot, is then compared against the position of the blip on the radar screen, and the flight positively identified. The aircraft can also be asked to make a turn onto a specific heading for identification, and this can be confirmed on the radar screen as the blip changes direction. Many areas of the world still use this type of radar.

Secondary surveillance radar
SSR is a more sophisticated form of radar, and operates using transponders on board the aircraft, as described earlier. Each aircraft is issued with a code by ATC which the pilot dials on the transponder and which identifies the aircraft on the radar screen. The transmission of this code is called 'squawking' and on initial contact the pilot is instructed by the controller to 'squawk' a certain code, e.g. 'Squawk code A 1133'. An identification button on the transponder

Radar Antenna.

can also be pressed by the pilot, when requested, to identify the aircraft positively by enlarging the blip on the radar screen. If identification is required, the word 'ident' is added to the code. The pilot then selects A 1133 on the transponder and presses the 'ident' button. Certain codes are memorized by pilots and used in the case of an emergency, complete radio failure, or a hijack. There is also a facility on the transponder to transmit aircraft height, which appears as a number by the relevant blip on the radar screen. It is used extensively in the USA and Europe, but most of the rest of the world does not as yet have the necessary ground equipment.

Precision approach radar (PAR)

PAR is used to monitor aircraft on the instrument landing system (ILS) during periods of bad weather, and also, in spite of the name, to monitor departures in similar circumstances. Since the equipment is seldom required and is expensive, it tends only to be found at the more difficult airports such as Hong Kong, where in bad weather PAR is used to monitor approaches to the north-west runway, and departures to the south-east as aircraft fly at low altitude between the outlying islands.

Ground control approach (GCA)
GCA is more often used by the military. A pilot is given instructions on direction and rate of descent by a radar controller, to guide the aircraft on a normal approach to a runway in instrument conditions, without the aid of instrument landing equipment. It is used occasionally at civil airports when the instrument landing system is unserviceable.

Radar control
There are different attitudes to the use of radar in the various parts of the world. In almost all the Third World countries, *en route* radar is non-existent because of the cost, and many airports are without radar. In Europe and countries like Australia, Canada and New Zealand, radar tends to be used as a monitor of the flight's progress, rather than to direct aircraft, except on the approach phase of the flight. Aircraft normally adhere to allocated departure routeings and closely follow flight plan routes while monitored by radar, except on occasions when radar headings are issued if a re-routeing is required.

In the USA it is quite different, with the emphasis very much on radar control. Departure routeings and flight plan routes are allocated as before (in case of complete radio failure when the aircraft follows the flight plan route), but as soon as the aircraft is in the air radar control takes over, and the aircraft is directed by radar headings along route. Sometimes complete flights over US airspace are conducted without following the flight plan route, the aircraft being continuously issued headings to steer by radar control, with the pilot changing from frequency to frequency as the flight passes from one radar controller to another along the way. Radar control of this kind allows direct routeings over long distances instead of aircraft following each other in line from beacon to beacon along an airway. Aircraft can be navigated automatically to a point many hundreds of miles away by using inertial navigation equipment, saving both time and fuel. A favourite direct routeing requested by pilots after a westbound Atlantic crossing is from the Labrador coast of Canada down to Kennebunk (250 n.m. north of New York), a distance of some 700 n.m. Direct routeings over such distances are rare in Europe or elsewhere.

The USA has a great deal of light aircraft traffic, and the controller often has in view on his screen a number of unidentified blips. The controller calls the pilot's attention to any traffic in the vicinity by indicating the positions of aircraft, relative to the pilot, using the hours of the clock, 12 clock being dead ahead. e.g. 'Traffic ten clock, range three miles, height unknown'. Because of the intensity of light aircraft traffic at lower levels, the airspace within the vicinity of airports has been nicknamed 'Indian territory' by airline pilots.

Solar activity

Intense solar activity can, on occasions, seriously affect the performance of radio equipment. Communications, navigation and radar systems are all susceptible to temporary interruptions. HF communications are particularly affected, although VHF transmissions can also suffer loss of signal. On the rare instance total radio blackouts lasting several hours can occur, and radio navigation equipment, such as Omega in the VLF band, can be rendered temporarily useless. Even ground telephone lines can be affected by interference.

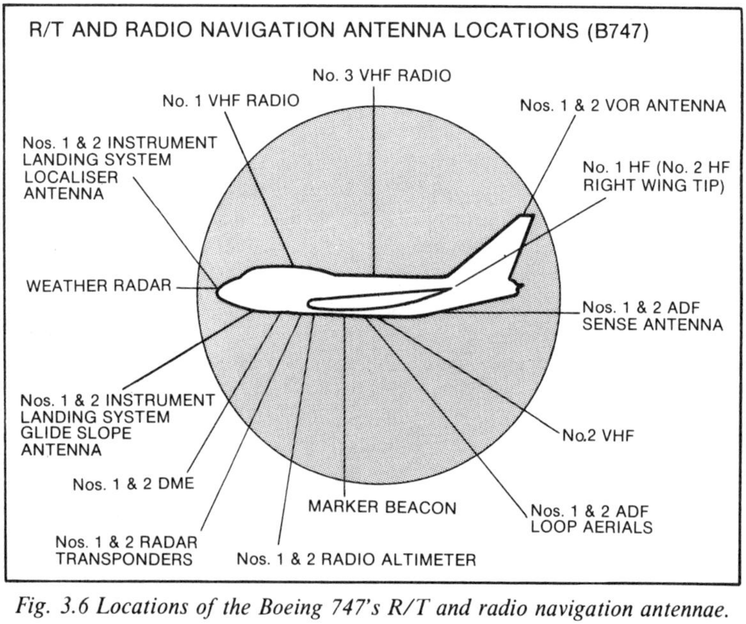

Fig. 3.6 Locations of the Boeing 747's R/T and radio navigation antennae.

Solar disturbances are related to sunspot activity with an eleven-year cycle that last reached a peak in late 1979 — believed to be among the most active in the last hundred years. A typical disturbance begins with the sudden eruption of a solar flare reaching many miles out into space. The flare emits a burst of X-ray and ultraviolet radiation and a stream of energy particles known as the solar wind. The radiation reaches the earth within a few minutes, and after about 20 minutes (and for up to 20 hours), high energy particles strike the earth's atmosphere. The atmosphere is deepest at the Equator and shallowest at the Poles where the air is colder and more dense. Particles penetrating the atmosphere are burned up high in space, except near the Poles where they penetrate quite deeply causing tremendous ionization in the lower levels of the Polar region ionosphere. The disturbance can last for several days.

After 20 hours (and up to 72 hours later), low to medium energy particles (protons and electrons) also penetrate the earth's atmosphere at the Poles, causing intense ionospheric disturbance and affecting the earth's magnetic field. These particles finally burn up in the lower atmosphere producing the dancing lights known at the North Pole as the Northern Lights, or Aurora Borealis, and at the South Pole as the Southern Lights, or Aurora Australis.

Chapter 4
Navigation – 1

From the voyages of the Phoenicians to the explorations of Captain Cook the great oceans of the world have yielded to man's ingenuity, witnessing outstanding feats of navigation in the most primitive of conditions. Instruments, charts, and clocks were unknown to the very earliest of travellers, but by some sixth sense they could smell the weather and read the sky like no man today. Advances in technology have brought an accuracy to navigation unknown in the past; but lost forever is that closeness to nature where men knew by instinct the waxing and waning of the moon, and could feel midday by a glance at the sun. Nowadays, in this computerized world, simple watches of extreme accuracy are commonplace, and equipment which sent men to the moon, the inertial navigation system (INS), is common on all the big jets.

In civil aviation, developments in the last decade have all but seen the demise of the navigator, although in the airforces of the world his skills are still required on many aircraft types. So navigators, although few in number, are still with us, working busily at small cramped desks, employing the same basic techniques that have been in use for over two hundred years — but now they are travelling one hundred times faster.

Although a number of short and long range radio aids are used by the navigator, the sextant, too, is still available when required. A position is obtained from the stars by pushing the sextant up through a hole in the flight deck roof, the sextant being preset to find the required star. Over such areas as the vast Sahara desert in North Africa the navigator has only the stars to aid him.

Within a comparatively short period the navigator will be gone forever in aviation. At the moment, for example, no aircraft (military or civil) is allowed within the North Atlantic track system unless equipped with the requisite navigational equipment, a pattern which will eventually be repeated throughout the world. Elsewhere, all long range aircraft must have INS or similar equipment on board, or carry a navigator, if the flight is in excess of 500 nautical miles over sea, desert, or other sparsely populated areas. Today the greater accuracy of computer navigation is required, not so much to navigate from one place to another, but to maintain precise lateral separation between aircraft in remote areas without radar, in an ever more crowded sky.

The inertial navigation system (INS) is one of the greatest breakthroughs in recent aviation history, replacing at a stroke the navigator whose knowledge and skills have been acquired over centuries of travel. However, although flight navigators are no longer carried on the big jets, descriptions of navigational practices and equipment have been included in this book because not only will the detail be found interesting, but the knowledge will prove useful in understanding explanations and information given in other chapters. A short history of the development of navigation is also included.

History

Navigation, in its infancy, developed at varying times in different parts of the world, although early techniques tended to follow similar lines. In the South Pacific, for example, navigation methods used were little different from those of early seafaring peoples of the Mediterranean but, unlike in the West, they remained relatively unchanged over the centuries.

Man's arrival in the South Pacific has been hailed as a navigational feat in itself, although whether by skill or accident is open to argument. It is known, however, that Polynesian and Micronesian navigators journeyed many hundreds of miles, perhaps even as far as Hawaii, by means of star courses passed down by word of mouth from generation to generation. The details memorized for each of the many journeys undertaken were quite considerable, and their knowledge of the movement of the sun and the stars was extremely good. They were not only aware of the basic techniques of navigation, but knew how to apply the complex corrections required to navigate with great accuracy.

On northern journeys, the Pole Star, stationary in the sky, was their indication of north, easily identifiable by the Great Bear (the Plough or Big Dipper). In the southern hemisphere, south was obtained from the Southern Cross and its pointers. They were aware that on the Equator the stars rise and set at the same points in the horizon throughout the year, but that the sun's point of rising varies, and that when viewed from north or south the stars have an apparent motion towards the Equator.

To the South Pacific navigator the sky was like one giant star compass. Not only would they employ basic guide stars, but they would also use different guide stars depending on the strength of the current. Some routes even had guide stars which allowed for leeway angle, the difference between the canoe's course and wake, caused by the wind effect on the sails. *En route,* they could not only divert to another island, if necessary, but could make direct for the harbour entrance. In winter the night sky is different from that in summer, requiring a complete new range of guide stars to be memorized by the navigator for each journey.

Half the journey time on any voyage would, of course, be in daylight and an alternative source of navigation was required. The sun was a primary aid; a course being maintained by observing the shadow of the mast and applying the appropriate corrections. A knowledge of local winds was also used as a secondary aid, but a major aid during daylight hours, and a primary aid during times of complete cloud cover when neither sun nor stars were visible, were the ocean swells which could be detected by the navigator from the movement of the boat. Waves are a result of a local wind blowing over the surface of the sea but swells are generated by strong prevailing winds, such as the Trade Winds, and maintain their direction and speed over long periods, persisting even hundreds of miles from the winds that produce them. The navigators could detect particular swells from the local wave pattern and could then use them as a heading reference. Great skill would be required by the navigator to differentiate between the motion of the primary swell and the movement of the boat, caused by, perhaps, a number of local wave patterns in a choppy sea. The technique required years of practice, and it was said that many navigators sensed the motion of the swell through their testicles, the most sensitive part of the anatomy available! What you might call navigating by the seat of one's pants!

So expert at the art of navigation were these men that one particular hundred-mile journey to a tiny island would be undertaken at the drop of a hat, at any time during the day or night, even when blind drunk on palm toddy.

In the Mediterranean, stars, local winds and landmarks were also the primary source of direction finding for the early seafarers, but at a very early stage implements of one kind or another were introduced as an aid to navigation. As early as the Egyptian period, the lead line for measuring depth, together with the sounding rod for shallower waters, were in common use. Written sailing instructions of some kind were probably also available. The lead line, however, with the later addition of a small scoop, was the first true navigational aid employed. As the depth was sounded, the scoop would lift a small portion of the seabed which would indicate the type of shore, and hopefully the locality. The first attempt at finding position by a navigational tool was now established.

At some time in the distant past man learned how to steer by the stars, but it was the Phoenicians, the long distance sailors of their day, who first discovered how to use the stars for navigational purposes. For almost 2000 years before the birth of Christ they plied their trade throughout the length and breadth of the Mediterranean, and even voyaged out through the Straits of Gibraltar, the 'Ocean Gate' into the wide Atlantic beyond.

Little is known of their navigational methods. With their ability they monopolized the distant trade routes, accumulating riches and wealth, and their navigational skills were, not surprisingly, a closely guarded secret. It wasn't until as late as AD 63 - 65 that written evidence existed that man used the stars in an attempt to determine position rather than simply to indicate direction. The height of the Pole Star, 'the never-setting Axis, brightest star in the twin Bears', was mentioned by the poet Lucan as being used by Roman seamen as an aid to navigation.

It is not impossible that the Phoenicians reached Britain and perhaps even further north. It is more likely that they reached the Azores, and perhaps, if only by accident, the coast of America, although no proof exisits that they ever reached any of these shores. Not until the Vikings, almost nine hundred years later, could the same claims be made of another seafaring people.

The navigational arts changed little in the thousand years after Christ. The Vikings ventured out across the seas using much the same principles of navigation as the Phoenicians before them, with only minor improvements. They left evidence that they knew of the changing declination of the sun, and are known to have used crude instruments to measure its height. In small open boats, staying as close to shore as possible, they voyaged into the Mediterranean and Adriatic, and over the Atlantic to Iceland and Greenland, and almost certainly to America.

In the 200 years following the Vikings, the introduction of the chart and the use of the compass were to revolutionize navigation. It is uncertain whether it was the Vikings or the Arabs who were the first to use the principle of magnetic attraction, and it is not known when the use of the compass was established. For a long time the lodestone was known to have the property of being able to magentize a piece of soft iron for a short period and thus, when pivoted, point due north and south. To the ordinary seamen of the middle ages this was undoubtedly witchcraft, to be treated with the greatest suspicion, and used as an aid only when all else was unavailable.

Early attempts were made with the soft iron needle placed on a piece of wood and floated in a tub of water, but this could be used at sea only in the calmest of conditions. The lodestone would be waved over the iron to temporarily magnetize the metal, and the needle would then take up a north - south direction on coming to rest. Later the needle was pivoted, and in this form, the compass and lodestone became familiar tools of the navigator's trade. The original wind rose of the Greeks showed the eight points of the principal winds, and the first mariners' compass adopted a similar layout, eventually being subdivided into 32 and then 64 points. By late in the fifteenth century the needle and rose were pivoted together, and the compass enclosed in a box. The box was then aligned with the fore and aft axis of the ship and a course could be read against the lubber line mark. Early in the sixteenth century the compass was set on gimbals, but as yet no account was taken of variation, the difference between true and magnetic north.

Maps appeared around the second half of the thirteenth century, but were more like informative lists than the detailed charts of today. Known as Rutters from the French 'routier', they contained details on tides, depths, landmarks and bearings, as well as a host of other useful information. Such knowledge was not available to all who sailed the seas, and the Rutters were closely guarded by the pilots who owned them. Compass roses were then marked on the charts for ease in measuring rhumb line courses (tracks cutting lines of longitude at equal angles), but not until Mercator's work in the latter half of the sixteenth century were charts drawn with accurate projections on which rhumb lines could be measured as true courses.

Originally the Pole Star and later the midday sun were used to measure latitude, the altitude (height of the body above the horizon in degrees) being taken roughly against the mast or similar structure. By the fifteenth century the astronomer's instruments of astrolabe, quadrant and cross-staff were modified for use at sea. The astrolabe, used by astronomers for hundreds of years, followed by the quadrant, each employed the plumb-line principle, and were extremely difficult to use on board, the navigators sometimes having to go ashore to take a reading. The cross-staff (Fig. 4.1) replaced both, being the simplest to make and easiest to use, and by the end of the sixteenth century was refined to form the back or fore staff which remained in use for over 300 years, until superseded by the sextant.

'Dead' reckoning was still used to estimate longitude, with the course being read from a compass, and time measured by the cabin boy repeatedly turning a sand glass. Accurate recording of speed, however, remained difficult. At first a large piece of wood or a log was thrown from the bows and the time it took to travel the ship's length was measured. Known then as the Dutchman's log, the word was adopted into the seafaring language and gave its name to the ship's log for measuring distance and speed, still in use today. Later a

The Deſcription of the Croſs-Staff.

This Inſtrument is of ſome antiquity in Navigation, and is commonly uſed at Sea, to take the Altitude of the Sun or Stars, which it performs with ſufficient exactneſs, eſpecially if it be leſs then 60 degrees, but if it exceed 60, it is not ſo certain, by reaſon of the length of the Croſs, and the ſmallneſs of the graduations on the Staff.

Fig. 4.1 The cross-staff.

piece of cord was attached and the log streamed astern until outside the effect of the ship's wake. A series of knots was tied at set intervals in the cord remaining and, when released, the number of knots passing over the stern was recorded in a certain time, thus giving a more accurate indication of speed. Today navigators still measure speed in knots, now equivalent to one nautical mile per hour.

In the 500 years after the introduction of the compass and chart, improvements in navigation were restricted to the refining of established practices. As man voyaged to the distant oceans, the difficulty of obtaining longitude was overcome by a ship first setting sail for the required latitude, and then sailing along the latitude until the destination was reached. When sailing to a lonely island the problem became acute. If the estimated arrival time had elapsed, it was not known if the island had been passed unnoticed or if the estimated time was incorrect and the island still to come. Whether to keep going or turn back was an agonizing decision for the navigator to make, and many a man perished when he made the wrong one.

In the middle of the seventeenth century, the Royal Observatory at Greenwich, near London, was founded by Charles II, and the start of the eighteenth century saw the establishment in London of the Board of Longitude, with large sums of money being offered as reward for any means of solving the problem. The idea of obtaining longitude by carrying an accurate time-piece on board had for some time been considered possible, but such an instrument was not available until the appearance of the chronometer, with an amazing accuracy of one-tenth of a second per day, around the middle of the eighteenth century. The problem of longitude was solved. Astronomers and scientists working in the Greenwich Observatory established the meridian through Greenwich (i.e. the line of longitude joining North and South Poles through Greenwich) as the datum for measuring longitude. With the chronometer on board set to the time at Greenwich, local midday of the ship's position could be observed from the sun's altitude, and at the same time the Greenwich time noted. Since the earth rotates at an angular velocity of 15° per hour, the time difference in hours between Greenwich and local midday, multiplied by 15, would be the number of degrees separation from Greenwich and would establish the ship's longitude. To this day Greenwich Mean Time (G.M.T.) is still used by shipping and aviation as the standard time setting throughout the world, and the Greenwich Meridian (zero longitude), internationally adopted in 1884, is now marked at the Greenwich Observatory by a brass strip where visitors can stand astride the line with one foot in the east and one in the west.

In the first half of the eighteenth century a world chart of magnetic variation was produced and, later in the century, the octant, followed by the sextant with all its refinements, was brought into service. The compass was now marked in degrees, with the needle permanently magnetized and damped to give a steady reading.

By the latter half of the eighteenth century the instruments of navigation were now established and they remained more or less unchanged for over 200 years until radio aids were introduced during the Second World War. In the last ten years the invention of the inertial navigation system and its introduction into service have revolutionized modern navigational practices, and once again a new dimension in navigational techniques has evolved.

Time

The sun

Basically, time is relative to the sun. The rotational motion of the earth on its axis results in the period of one day. The movement of the earth in its orbit round the sun is a measure of one year. The earth rotates in a west to east direction, with the sun rising in the east and setting in the west, so times in the east are ahead of times to the west. Time does not change with north-south movement on a meridian. The axis of rotation of the earth is at an angle to its plane of motion round the sun (Fig. 4.2), and as the earth spins throughout the day the sun rises at dawn, appears to climb and descend in the sky, then

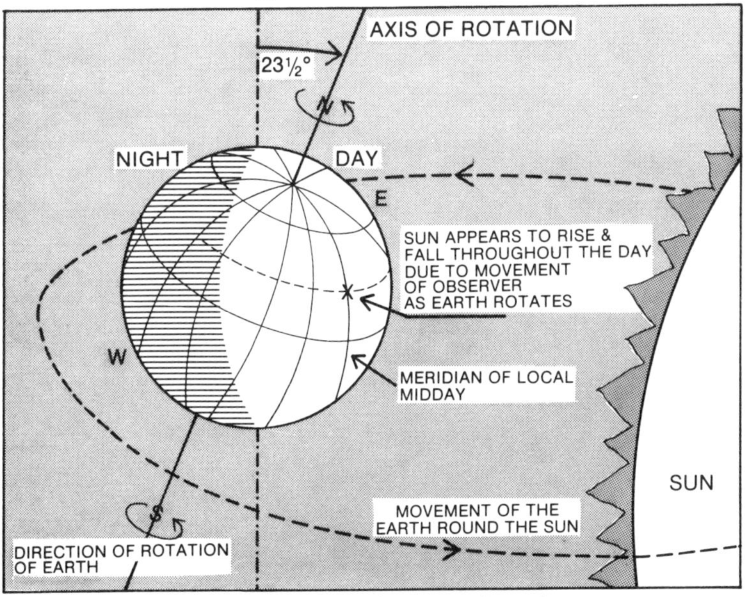

Fig. 4.2 The earth and its axis of rotation.

falls from view at dusk. Local midday occurs when the sun is positioned vertically above an observer's meridian, and the sun is seen to be at its highest point in the sky. The combined effects of the angle of the axis of rotation and the movement of the earth round the sun also give rise to the apparently higher sun in summer and lower in winter, and result in the seasonal changes of weather experienced throughout the year.

Local Mean Time (LMT)

Local mean time is the measurement of time at one position, and will only be the same for places on the same meridian. LMT in London is, therefore, a few minutes ahead of Cardiff, and likewise New York a few minutes ahead of Washington. To standardize time in one country, the time adopted is usually close to the LMT of the meridian through the capital of the country concerned (Fig. 4.3). For example, in the United Kingdom, the LMT at Greenwich (i.e. Greenwich Mean Time — GMT) is the standard time for the whole of the British

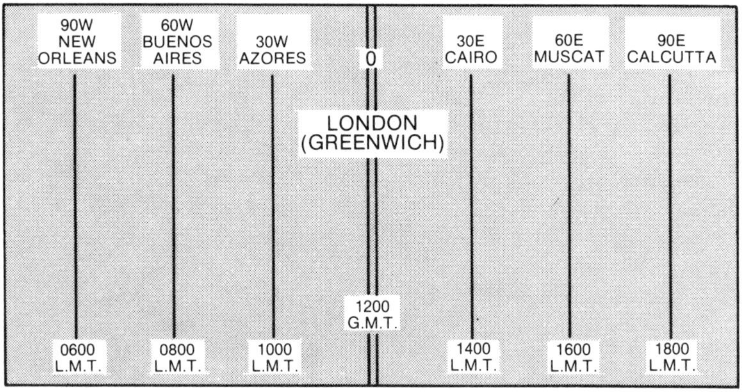

Fig. 4.3 Local mean times.

Isles, except during summer when one hour is added to GMT for daylight saving. Countries with a large east-west spread such as the USA, Canada, Russia and Australia, are broken into time zones across the country.

Since the earth is divided into 360° of longitude, 180° to the east and 180° to the west of Greenwich, and also rotates at 360° in 24 hours (15° per hour), a simple relationship exists between longitude and time. A difference in longitude between two places of 15° represents a time difference between the places of one hour. A time at a position three hours ahead of GMT establishes that the longitude of the position is 45°E (i.e. 45° east of the Greenwich Meridian).

Therefore, using the Greenwich Meridian as a datum and GMT as the world standard time, the longitude of a position can be used to establish the time difference in hours between the position and Greenwich by simply dividing the longitude by 15 (Fig. 4.4). And, vice versa, knowing the time difference, the longitude can be calculated.

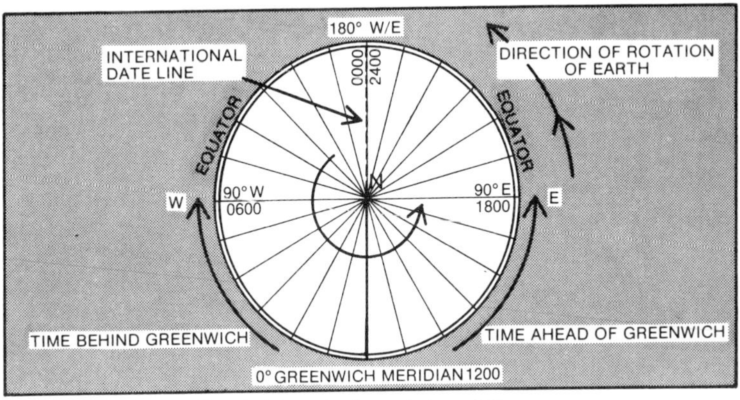

Fig. 4.4 Longitude and time difference.

Time zones

The earth is divided into time zones each 15° wide, with each zone one hour different from the next. In practice the zone boundaries are altered to coincide with the borders of countries, or rearranged to include areas such as island groups within the same time span.

International Date Line

As time is ahead of Greenwich by one hour for every 15° to the east, by 180°E (the antemeridian of Greenwich) the time will be 12 hours ahead. To the west, time is one hour behind Greenwich for every 15°, and so by 180°W (the same line as 180°E) time is 12 hours behind Greenwich. Therefore, at the same point on longitude 180° E/W, the time is both 12 hours ahead and 12 hours behind the Greenwich datum, i.e. at the same time but on different days. To overcome this problem the International Date Line was established. The line runs down the 180° meridian of longitude, but once again is altered to suit international boundaries. For example, when GMT is 1200 at midday Monday, it will be 2400, i.e. midnight Monday night, just to the east of the date line and 0000, i.e. midnight Sunday night, just to the west.

Crossing the date line at about this time from west to east, say from Honolulu to Fiji, a traveller would find himself flying a few miles to the west of the date line at about one minute to midnight Sunday, and two minutes later a few miles to the east of the date line at one minute past midnight Tuesday morning. Monday would be gone. Crossing the date line east to west at the same time, two minutes after one minute to midnight Monday, the traveller would find himself at one minute past midnight Monday morning, and would have to face Monday all over again. Fine if it was Christmas Day and two could be enjoyed in one year, but a disaster for many if it was New Year's Eve going the other way, and the celebrations were missed.

Daylight saving

During the First World War the Government in the UK surmised that production could be increased by keeping people out of the pubs during the hours of the working day and, in summer, by getting them up earlier to take advantage of the extra hours of daylight. Action was taken by closing the pubs in the afternoon, and by advancing the clocks one hour in the summer. Workers who started at 8 a.m. continued to start at the same time by the clock in the summer, although it was in fact still only 7 a.m., and a whole nation was coaxed from bed early by the simple expedient of putting the clocks forward one hour. Daylight saving, as it is now known, is still with us, and has been adopted by many countries throughout the world, the clocks being advanced one hour in the spring and retarded one hour in the autumn. It is often confusing which way the clock has to be moved at which time of year, and in the United States Americans remember the sequence as: Spring forward; Fall back.

Time signals

In the days of sail, an observatory was positioned to be visible from the harbour. Each day at one o'clock, a large ball atop a mast on the observatory roof was dropped, the precise time being indicated by the end of its fall. Where the positioning of an observatory near the harbour was not possible, a gun was fired at one o'clock. Ideally midday was the hour to give the time signal, but as the astronomer was making his observations of the sun at this time, the signal was given one hour later. The tradition is still carried on in places as far apart as Edinburgh in Scotland, where a gun is fired daily, and in Sydney, Australia, where the observatory ball is dropped each weekday at 1 p.m. (Fig. 4.5).

Fig. 4.5 Sydney Observatory.

Time signals now are transmitted throughout the world from various sources. The BBC transmits worldwide, on the hour, a six-pip signal at one-second intervals, the sixth and longest pip being the precise time. At Fort Collins, Colorado, and in Hawaii, radio stations fed from atomic clocks with an accuracy of 1 sec/million years, transmit continuous time signals throughout the world on 2.5, 5.0, 10.0 and 15.0 MHz. A pip is transmitted each second, and every minute a voice statement of the time is made and is followed by a tone indicating the precise minute. 'At the tone it will be thirteen forty-four co-ordinated universal time'. The time being transmitted in all cases is Greenwich Mean Time, which has been accepted as the world standard time. Since Fort Collins (Station code WWV) and Hawaii (Station code WWVH) transmit on the same frequencies, time statements are made by a male voice from Colorado and a female voice from Hawaii.

Leap second

Greenwich Mean Time is the basis of Co-ordinated Universal Time, and since 1 January 1972 has been related to International Atomic Time (IAT), which is time as maintained by an atomic clock. Co-ordinated Universal Time was assumed to flow uniformly, but the more accurate time-keeping equipment that has become available has shown that this is not so. Because of the tides, winds, and seasonal variations, the length of the average day fluctuates by about a millisecond, and the length of the average day is increasing by about 2 milliseconds per century because of tidal friction. Since IAT is the time reference used today, the irregularities in the earth's rotational motion mentioned above result in the two times becoming out of phase after some years, and GMT and IAT have to be re-aligned periodically by applying a correction in the GMT scale. One second, known as a 'leap second', is inserted or omitted on a particular day, the day being chosen by the International Time Bureau. In 1979 the final minute of the year was extended to 61 seconds, and the main time centres throughout the world adjusted simultaneously at midnight GMT on New Year's Eve; the BBC for example, transmitting a seven-pip time signal instead of the usual six to accommodate the extra second.

Measurement of day

Figure 4.6 is a plan-view diagram (not to scale) of the earth's plane of movement round the sun. If the earth remained stationary rather than orbiting the sun, then after one daily revolution of exactly 360° the sun would be vertically above the meridian through point A. However, as the earth does move in its orbit round the sun, to

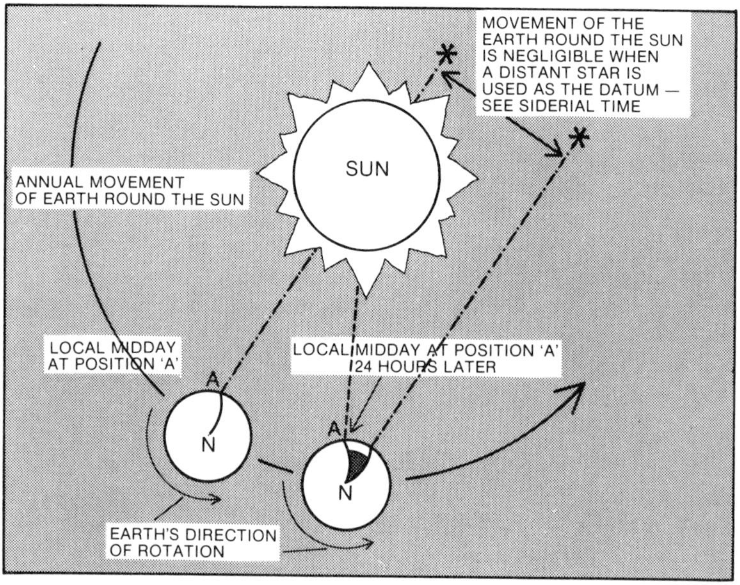

Fig. 4.6 The earth's plane of movement round the sun.

complete a period of one day the earth has to revolve a little more than 360°, as indicated by the shaded angle (in fact about 361°), before the sun is once again vertically above the meridian through point A. Therefore, the measurement of day not only depends on the revolution of the earth on its axis, but also on the movement of the earth in its orbit round the sun, which is assumed to be constant. The speed at which the earth orbits the sun in fact varies, and so the actual day is normally 24 hours plus or minus a few minutes. The 24-hour period is an average of all the days throughout the year. Since an astronomer observing the sun obtains a time at midday relevant to the sun of that particular day, a correction has to be applied to convert the observed time to the average time based on 24 hours. The correction involves two factors relating the angle of the axis of the earth's rotation to its plane of motion round the sun, and to the variable speed of the earth in its orbit round the sun. The correction is normally presented in the form of an equation known as the 'equation of time.'

Siderial time

When the rotation of the earth is observed in relation to a datum such as a star, the distance of the earth from the star is so great (the nearest star is more than four light-years away) that the orbit of the earth round the sun can be ignored, and the earth can be assumed to rotate in a stationary position. In this case the earth rotates exactly 360° in relation to the star, and the day relevant to the star is measured as 23 hours 56 minutes 04 seconds. This time relative to the stars is known as siderial time, and is the basis of time used by astronomers.

The year

The average length of the calendar year is 365.2425 days, which is generally accepted as 365¼, 365 being the standard year, with an extra day added each leap year (i.e. when divisible by 4). However, each year the fraction of a day over 365 is, as can be seen, just less than a ¼, and an allowance for this has to be made by having the standard 365 days each century year, unless it is divisible by 400 when it is counted as a leap year. Thus 1900 (although divisible by 4) was not a leap year, but 2000 will be a leap year (divisible by 400).

Seasons

Since the rotating earth acts like a giant spinning top, it maintains the properties of a gyro and the axis of rotation remains fixed in space as the earth orbits the sun. In fact, with the earth rotating at 360°/24 hours, the speed on the surface of the earth at the Equator is just over 1000 miles per hour. This speed on the surface reduces towards the Poles as the effective radius decreases, and is one reason why space flights are launched from near the Equator, the higher surface speed helping to throw the space ship away from earth and out of its field of gravitational influence.

In the northern summer, the northern hemisphere is tilted towards the sun, and the surface is heated more uniformly, the opposite being true in winter. In the southern hemisphere the seasons are directly opposite to those in the north, but the solstices and equinoxes are based on the seasons of the northern hemisphere (Fig. 4.7).

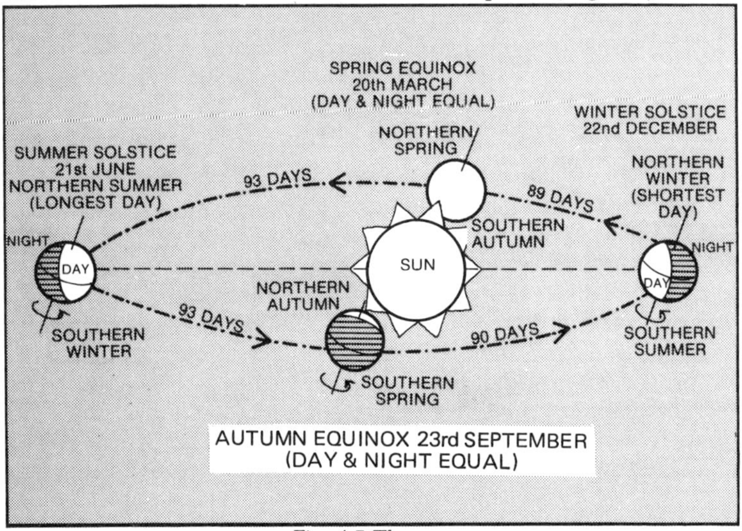

Fig. 4.7 The seasons.

The plane of the Equator is inclined at an angle of 23½° to the plane of the earth's orbit round the sun. As a result, in the course of a year (i.e. one complete orbit) the sun appears to shift from latitude 23½ N to 23½ S. When the sun lies at its furthest north on about 21 June, it is said to be the summer solstice, the longest daylight, and latitude 23½ N is named the Tropic of Cancer. When it is at its furthest south on about 22 December, it is at the winter solstice, the shortest daylight, and latitude 23½ S is named the Tropic of Capricorn. Between the Tropic of Cancer and the Tropic of Capricorn the sun is more overhead, and therefore at its most effective. This area is, of course, known as the tropics. When the path of the sun crosses the Equator from south to north on about 20 March it is known as the spring equinox, and in the opposite direction on about 23 September it is known as the autumn equinox.

Length of day

The shortest day at winter solstice is the effect of the northern hemisphere being tilted into the shadow of night, resulting in perpetual darkness at the North Pole throughout the winter months. At summer solstice, the northern hemisphere is tilted into the area of daylight, giving the longest day, and resulting in perpetual daylight at the North Pole throughout the summer months. The opposite is true at the South Pole.

Arctic Circle

The Arctic Circle defines the extremity of perpetual darkness in winter and perpetual daylight in summer at the North Pole. The Arctic Circle is therefore at Latitude (90 -23½) = 66½ N. The Antarctic Circle serves a similar purpose at the South Pole at 66½ S.

Jet lag

In mid-summer in Europe, the spawning of certain oysters takes place during the spring tide, which always follows a full or new moon. If these oysters are moved to similar surroundings on the other side of the world at about this time, the oysters become confused, and the spawning process is upset because of the differing tidal patterns. During this period of confusion the oysters could be said to be jet lagged.

The human body, too, becomes confused when the normal daily routine is disrupted, and it can take between one week and ten days to overcome the effects of jet lag, depending on the time change and the general condition of the passenger. The body can usually cope with up to three or four hours' time change without too much difficulty, but problems occur with five hours' time change or more. The body conforms to a pattern of events throughout a 24-hour period known as the circadian rhythm. This rhythm alerts the body to the time of day by passing the required signals at the right time: hunger when it's time to eat, waking when it's time to get up, etc. It also slows down the bodily functions through the night, during the hours the body is normally sleeping. When a person is moved quickly from one time zone to another, it takes a while for the circadian rhythm to adjust to the new sequence and jet lag is experienced.

Going west, let's say from London to New York, there is a five-hour time change, with New York five hours behind. By 8 p.m. New York time, it is 1 a.m. London time, and the body is feeling tired. To overcome the time change within a reasonable period, it's best to try and stay up until a near normal bedtime by the New York clock. If you retire at about 10 p.m. New York time, it will then be 3 a.m. London time, and the body will be very tired, but there's a chance of sleeping through until 7 a.m. or 8 a.m. the next morning, and thereby adjusting to local time relatively painlessly.

Going east is not so simple. From London to Bombay, for example with a time change of 5½ hours ahead of London, trying to go to bed at a normal time by the Bombay clock is not easy, as it will still be afternoon in London, and difficult to sleep. It is better to go to bed late when tired, and then force oneself up at the normal time, even if only after a few hours' sleep. You may feel jaded on the first day, but you should be able to sleep at the normal time that night and be reasonably adjusted to local time the following day.

When the transition is from one side of the world to the other, the problems are compounded. One wakes early in the morning ravenously hungry, the circadian rhythm signalling early evening and time for dinner. During the day the circadian rhythm thinks it's night, and the bodily functions are slowed. The traveller feels weak and confused, and sometimes a little dizzy.

How one person copes with jet lag compared to another depends on so many different factors, and there is no really quick solution to the problem. A person's condition on arrival can influence the effect of jet lag; the fitter and healthier the traveller, the easier jet lag is to overcome. A fit, healthy youth will find it easiest to adjust. A passenger should eat carefully, drink little or no alcohol (preferably a modicum of wine) but plenty of a variety of liquids (avoiding large quantities of any one drink), cut out smoking, and sleep when able. A little exercise walking slowly round the aircraft, or simple isometric exercises in the seat, will also help. An overnight flight, however, can play havoc with any such system.

It is said that either sex, drugs or drink can be an aid to overcoming jet lag. Since the last of the three is more readily available to most travellers, a few drinks on arrival to keep one awake or put one to sleep will certainly do no harm. In general, the best results are obtained by attempting to adjust to local time as painlessly as possible within the first two or three days and involves staying up to adjust going west, forcing oneself up in the morning when going east, and preparing for the worst when travelling to the other side of the world. Swimming is an excellent form of after-flight exercise.

People who travel regularly, like aircrew, learn to cope with being jet lagged, rather than being able to overcome jet lag more easily. Also, legislation such as 'Flight Time Limitations' is applied at all times to contain duty days within a reasonable period, and to ensure that adequate rest periods are available, when required, to reduce fatigue as much as possible. On the flight deck, pilots who normally check something twice, when feeling jet lagged will check items several times, where necessary, to ensure that correct procedures are maintained. They just know they are not functioning as they should and instinctively exercise greater caution. Businessmen, too, face the same problems, and are advised not to make major decisions for at least a day or two after experiencing a large time change.

Charts

Types of charts

Airways charts contain a mass of details on runways, radio beacons, communication frequencies, etc., and are used by pilots to check the aircraft's progress, whether on long range flights, airways flying, or on the approach on take-off phase in the local area of an airport. Navigation charts are used for plotting position by the navigator. Topographical charts are usually carried, as a safety requirement, on all long-haul flights.

Chart requirements

Representing the earth on a flat surface cannot be achieved without some distortion, and it follows that all flat maps suffer from such inconsistencies. The form of the earth is therefore represented on charts by displaying the earth's detail on a flat sheet in such a way that the *amount* and *kind* of distortion is known. The basic chart requirements for pilots and navigators alike is that angles, and therefore bearings, are correctly represented at all points on the chart. Charts designed with this property are known as conformal or orthomorphic, and this condition is achieved when meridians of longitude and parallels of latitude cross at right angles and when scale at any point changes at the same rate in all directions.

Chart construction

A geometric chart projection can be demonstrated by imagining a light source placed within a transparent earth, with the shadows of the earth's details being projected onto a sheet touching the surface of the earth. For the purpose of the projection the sheet may be a flat screen, a cylinder wrapped round the earth, or a cone sitting on top of the earth like a pointed hat, the cylinder or cone then being opened out to become a flat chart.

Three areas of the world, the Equator region, the mid latitudes and the Poles, each pose their own problems when representing the earth's surface as a flat sheet. In the Equator region the meridians are almost parallel to one another, in mid latitudes more inclined towards each other, and at the Pole they radiate outwards like the spokes of a wheel (Fig 4.8). Many different types of projection have been developed to overcome the problems associated with each area, but the three most commonly used are named Mercator, Lambert and Polar stereographic.

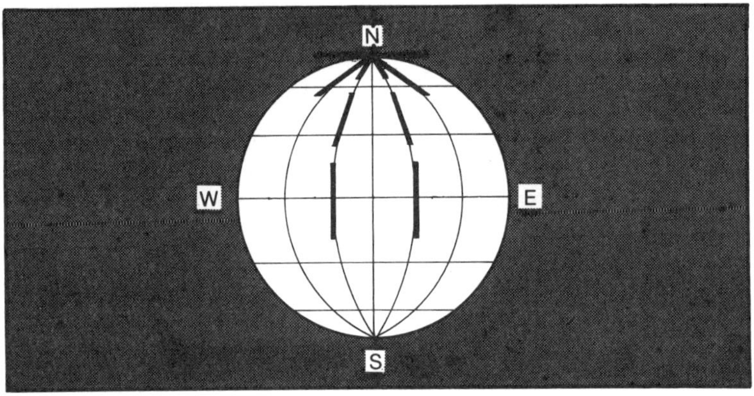

Fig. 4.8 Convergence of meridians.

Mercator Geometric Cylinder

In 1569 Mercator developed this chart for rhumb line navigation. It was the first conformal projection where true bearings could be simply and accurately plotted, and rhumb lines drawn as straight lines on a chart. Since meridians on the charts are drawn parallel, the earth is correctly represented at the Equator only. The scale expands with increasing latitude, resulting in expanded land forms. On a Mercator map of the world, chart distortion results in Greenland, for example, appearing on the chart to be much larger than is shown on a globe. The Mercator chart is normally more useful in the lower latitudes.

Lambert's Conformal Conic

Lambert, another famous cartographer, later developed a projection whereby great circles are drawn as straight lines on the chart. The earth is correctly represented at only two chosen latitudes, known as standard parallels. This chart is normally used in mid latitudes where the projection more correctly represents the shape of the earth (Fig. 4.9).

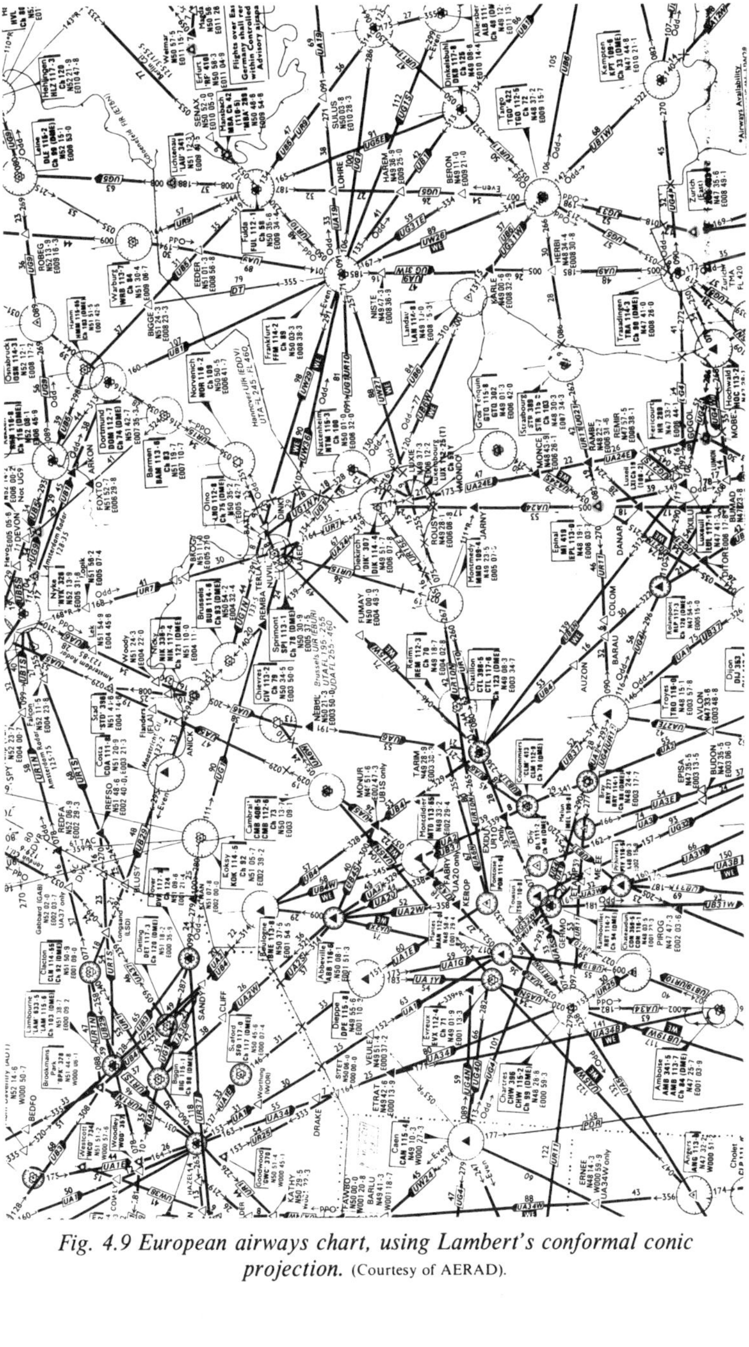

Fig. 4.9 European airways chart, using Lambert's conformal conic projection. (Courtesy of AERAD).

Polar Stereographic

Great circles are also drawn as straight lines on this projection. The earth is correctly represented only at the Pole. This chart is used for Polar navigation and trans-Polar flights.

Scale

On navigational charts the unit of one nautical mile is used and is equivalent to one minute of latitude. The scale at any point on a chart is therefore represented by the graduations of minutes of latitude on a meridian near that point. The distance between two points is measured by dividers and read from the meridian near the points.

Chart accuracy

Early charts were inaccurate and poorly drawn and frequently showed islands and even large land masses in places where they were not to be found. Before the Dutch explorations, for example, it was believed by many that a vast ocean stretched from China and the East Indies to the shores of the American continent, now known as the Pacific. To the early chart makers it was incomprehensible that such an ocean could exist without the presence of land, and early charts bore the rough outline of a vast imaginary land in the South Pacific inscribed with the words 'terra incognita australis' — the unknown land in the south. A large land mass was, of course, eventually discovered in the south and named by the Dutch as 'New Holland', and latterly by the British as 'Australia', taking its name from the above inscription. The newly discovered continent, however, bore little resemblance to the outline of the early charts.

Chart making today is a precise science, but even on present charts, especially in the Polar regions and remote areas of the large continents, it is not unusual to find sections of the chart marked 'uncharted', or 'unsurveyed'.

Chapter 5
Navigation – 2

The flight navigator has now disappeared from the flight decks of all the major civil airline operators on long-haul routes. However, many details of past navigation procedures provide good background information for present computer navigational practices on the big jets, and a brief outline of the past duties of the flight navigator is included. As an introduction some definitions, measurements, and units are given below.

Simplified definitions

Equator. The Equator is an imaginary line on the earth's surface, lying in an east-west direction, dividing the earth north and south. The Equator is the datum for the measurement of latitude.

Greenwich Meridian. A meridian is an imaginary line on the earth's surface, lying in a north - south direction, joining the North and South Poles. The local meridian through a position joins north, the position, and south in line, as does the meridian through Greenwich. The Greenwich Meridian is the datum for the measurement of longitude.

Rhumb line. A rhumb line is a line cutting all the meridians at the same angle.

Great circle. A great circle is any imaginary circle on the surface of the earth whose centre is the centre of the earth. The shortest distance between two points on the earth's surface is along the great circle that passes through the points. The Equator is an example of a great circle and the Greenwich Meridian an example of a semi-great circle.

Latitude. The latitude of a position is the angular vertical displacement of the position north or south of the Equator, measured at the centre of the earth from 0° - 90°. Latitude is expressed in degrees (°) and minutes (′), there being 60 minutes in one degree.

Longitude. The longitude of a position is the angular horizontal displacement of the position east or west of the Greenwich Meridian, measured at the centre of the earth from 0° - 180°. Longitude is also expressed in degrees and minutes. The exact position of JFK airport at New York is N 40° 38.9′, W 073° 46.9′ (Fig. 5.1).

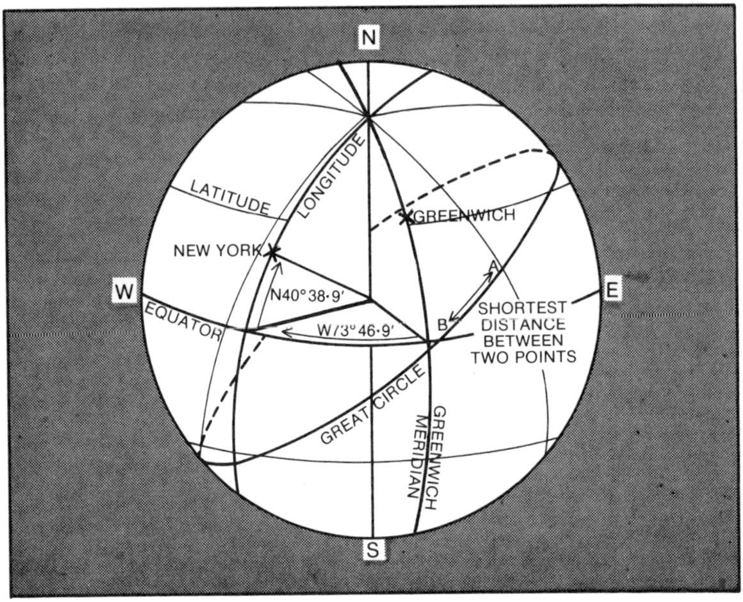

Fig. 5.1 Latitude, longitude and great circles.

Measurements

Bearing. A bearing is the direction of one point from another (Fig. 5.2).

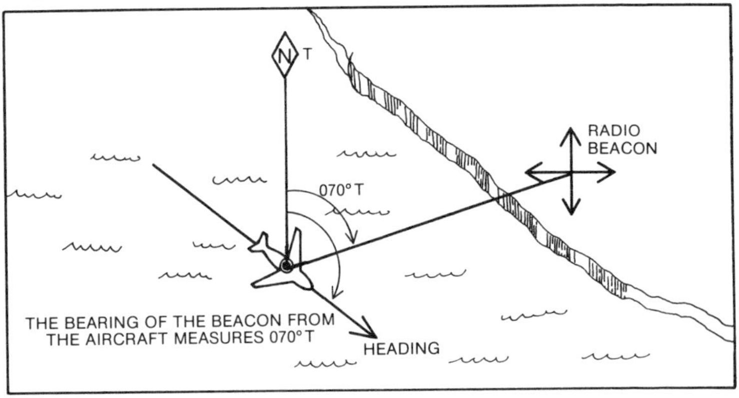

Fig. 5.2 Bearing.

Heading. The heading of an aircraft is the direction in which the nose is pointing.

Track. Track is the actual path of the aircraft over the ground.

Drift. Drift is the angle between heading and track required to correct for wind.

Direction. The direction of bearings, headings, tracks and winds are measured clockwise in degrees from North, and are expressed as a three-figure group from 000° to 360°. North is, therefore, 000° or 360°, East 090°, South 180°, and West 270°. True North (N_T) is the direction of the geographic North Pole, and directions measured from True North are expressed in degrees true (°T). Magnetic North (N_M) is the direction indicated by a pivoted magnetic needle influenced by the earth's magnetic field, and directions measured from Magnetic North are expressed in degrees magnetic (°M). The Magnetic North Pole moves slowly each year, and at the moment is positioned on the Canadian Island of Bathurst, 1000 miles from True North. Five thousand million years ago the Magnetic North Pole lay in the Eastern Pacific.

Variation and deviation

Variation is the angular difference between True and Magnetic North and, as the name implies, varies in value throughout the world. Dashed lines known as isogonals are drawn on charts joining points of equal variation. The variation experienced depends on the observer's position in relation to True and Magnetic North, and is named east or west depending on whether Magnetic North lies east or west of True North.

Deviation is caused by a local magnetic influence on board an aircraft resulting from the presence of ferromagnetic material or electrical circuits such as those operating radio equipment, which affect the compass and which, quite literally, divert the needle from Magnetic North. This artificial north indicated by the compass is known as Compass North (N_C) and directions measured by the compass are expressed in degrees compass (°C). Deviation, therefore, is the angular difference between Compass and Magnetic North and is named east or west depending on whether Compass North lies east or west of Magnetic North. The compass is arranged to maintain deviation at a minimum, and this process is known as swinging the compass. The remaining deviation is tabulated for various points of the compass and a card indicating the deviation corrections to be applied is placed on the aircraft.

Directions from charts are measured in degrees true since chart meridians point to True North. Chart directions, therefore, are converted to degrees magnetic by the application of variation, and degrees magnetic are converted to degrees compass by application of deviation. A true heading, for example, measured from a chart, required to navigate to a distant point, has to be corrected to steer by the compass. True heading (°T) ± variation (V) = magnetic heading (°M) ± deviation (D) = compass heading (°C). (Remembered by the mnemonic True Virgins Make Dull Companions.) Likewise, a compass bearing, say, from a distant radio beacon, has to be corrected for plotting on a chart. Compass bearing (°C) ± Deviation (D) = magnetic bearing (°M) ± variation (V) = true bearing (°T). (Remembered by the mnemonic Cadbury's Dairy Milk Very Tasty.)

Units

Nautical mile. One nautical mile (n.m.) is equivalent to the distance subtended by one minute of arc of a meridian on the earth's surface; i.e. one nautical mile is equal to one minute of latitude. The shape of the earth is, in fact, an oblate spheroid, being slightly flatter than a sphere at the top and bottom. One minute of arc at the North and South Poles is therefore different to one minute of arc at the Equator, so the average distance of one minute of arc, 6080 feet, is taken as the measurement of one nautical mile. (The Polar diameter is, in fact, only about 27 miles less than the Equatorial diameter, and for most navigational purposes the earth can be regarded as a true sphere.)

The **Statute mile** (s.m.) is equal to 5280 feet; 1609 metres. The nautical mile is, therefore, approximately 15 per cent longer than the Statute mile.

The **kilometre** is the length of one ten-thousandth part of the distance along the meridian through Paris between the Equator and the North Pole, and is equal to 3280 feet; 1000 metres.

Speed

Speed is measured in knots, one knot being equal to one nautical mile per hour. Ground speed (G/S) is the speed of the aircraft relative to the ground, and true air speed (TAS) is the speed of the aircraft relative to the air through which it is flying.

Wind velocity (W/V)

Wind velocity is the combination of the wind direction and speed. The wind direction always indicates the direction *from* where the wind is blowing. W/V is expressed as, for instance, 090/20 - the wind is blowing from 090° (East) at twenty knots. The wind direction is normally expressed in degrees true for upper winds and forecast surface winds, and degrees magnetic for actual surface winds. The wind is said to veer when its direction moves clockwise, and to back when its direction moves anticlockwise.

Navigating techniques

Basic navigation: 'Dead' reckoning and air plotting

The basic principle of navigation (by the navigator plotting on a chart) is to maintain a rough estimate of position, and from this knowledge and the equipment available, to obtain an exact pinpoint of position known as a fix. A knowledge of the approximate position of an aircraft was, at one time, maintained by a simple process known as deduced reckoning, which, when shortened to ded. reckoning, is known and pronounced by all navigators as 'dead' reckoning. An airplot of aircraft air position, using heading and true airspeed, but assuming no wind, was maintained throughout flight. Application of the forecast wind vector, at any time, established the estimated, or dead reckoning, position. This process is outlined below and shown in Fig. 5.3.

An aircraft setting off from overhead airport A and maintaining a heading of 090°T is subject to drift caused by the wind, and its track over the ground is different from its heading. In other words, the aircraft crabs in a slight sideways fashion across the surface of the earth. If a line is drawn on the chart in the direction of the aircraft heading, 090°T, and after, say, one hour, point B is marked on the heading vector using the true airspeed of the aircraft and the scale of the chart, then B is the position of the aircraft after one hour if there is no wind. The distance A - B now represents the true airspeed of the aircraft in nautical miles per hour (knots).

The forecast wind is now applied at point B. If, say, a wind of 045°/50 knots is used, then the line B - C is drawn to scale representing one hour's worth of wind blowing from 045°T at 50 knots. Point C is now the dead reckoning position after one hour. It is not the exact position, as the assumption has been made that the forecast wind has remained accurate at 045°T/50 over the past hour and this is seldom true. The measurement of line A - C gives the estimated ground speed of the aircraft in knots, and the direction of line A - C, 095°T, is the estimated track of the aircraft over the ground. The shaded angle represents the estimated drift of the aircraft, 5° to the right.

The triangle A B C is known as the vector triangle and from a knowledge of any two vectors the triangle can be drawn to scale,

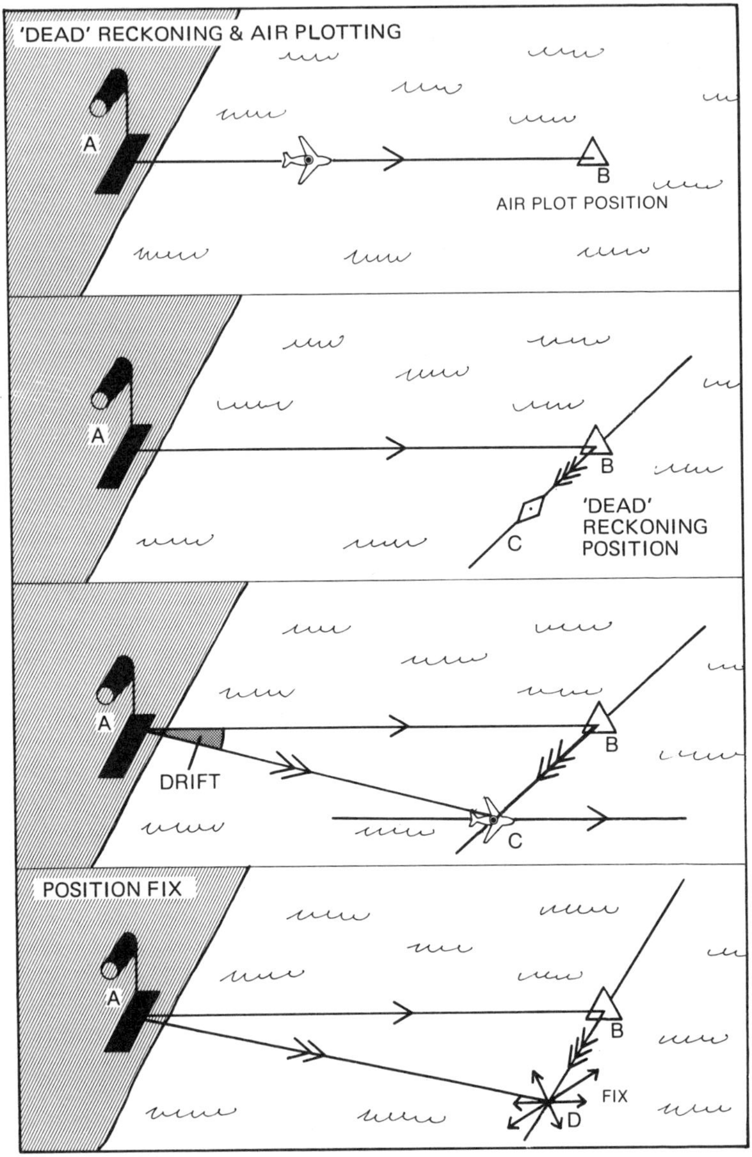

Fig. 5.3 'Dead' reckoning and air plotting.

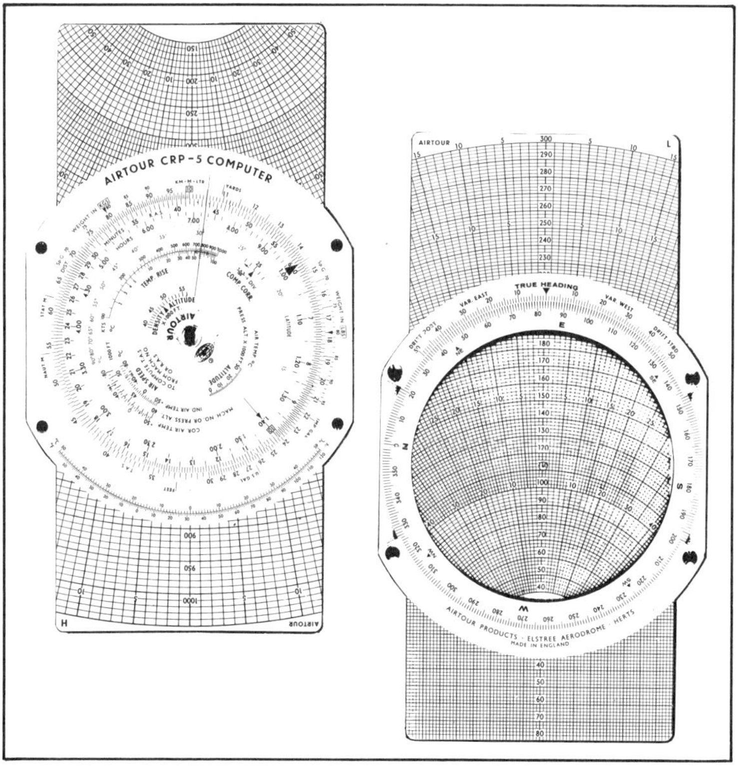

Fig. 5.4 The air computer.

either on a chart or on graph paper, and the third vector obtained. If, at the same time, a fix from independent sources establishes the aircraft position to be at point D, then vector A - D is the actual ground speed and track experienced since point A, and vector B - D the actual wind encountered. These calculations can then be used to correct heading and speed for the next leg of the flight.

Air plotting procedures were made redundant with the introduction of Doppler, a self-contained airborne navigation aid which transmitted radio signals towards the ground, forward and aft, from an antenna system on the underside of the aircraft. The frequency shift detected in received signals bounced back from the surface established the actual ground speed and drift. Armed with this information, and the true air speed easily obtained from graphs, the navigator could simply calculate the wind at any time using the air computer, which was named before the push button age, and is, in fact, a compass rose superimposed upon a sliding scaled graticule, displaying true airspeed and drift lines. The relevant section of the vector triangle was marked on the scaled graticule and the information obtained by rotating the compass rose, thus avoiding the necessity of drawing the triangle to scale on a chart. On the reverse is a circular slide rule which can be used for all arithmetic and aviation calculations. The air computer is still the basic tool of the flight navigator, and both sides of the instrument are shown in Fig. 5.4.

Standard navigation

The standard system of navigation used the Mercator chart, True North for plotting, and the magnetic compass for steering the aircraft. The Mercator chart was originally used to simplify navigation. Steering a constant magnetic heading (assuming constant variation and wind) resulted in a rhumb line track being flown, and could be simply plotted in true direction on this chart as a straight line. Since the meridians are projected parallel to each other the plotting of true direction was also made easier. The rhumb line track is longer than the great circle track, but since this system was used in the lower latitudes, the difference was minimal.

Grid navigation

In higher latitudes charts based on the normal Mercator projection become invalid because of the rapidly expanding scale and the inability to show the Poles. In mid and Polar latitudes, therefore, Lambert's conformal and Polar stereographic charts were used for navigation.

On such charts, however, the meridians are not parallel to each other, but point more towards the Poles as they would on the surface of the earth. On these charts a straight line marks a great circle track which cuts the meridians at different angles. To maintain a great circle track using standard navigation techniques required a heading expressed in degrees true to be changed continuously *en route* in the mid latitudes and rapidly near the Poles. To overcome this problem the grid navigation system was developed.

An arbitrary grid was superimposed on the chart, the direction of the grid being known as Grid North (NG), and was the datum for plotting bearings, expressed in degrees grid (°G). For convenience, the Greenwich Meridian was normally selected as Grid North, and the grid meridians were, quite simply, superimposed on the chart each parallel to Grid North (Fig. 5.5). In this manner, although the great circle track cut the true meridians at different angles, it cut all the grid meridians at the same angle. The heading of the aircraft could then be expressed relative to Grid North, and this constant grid heading applied to maintain the aircraft close to a great circle track.

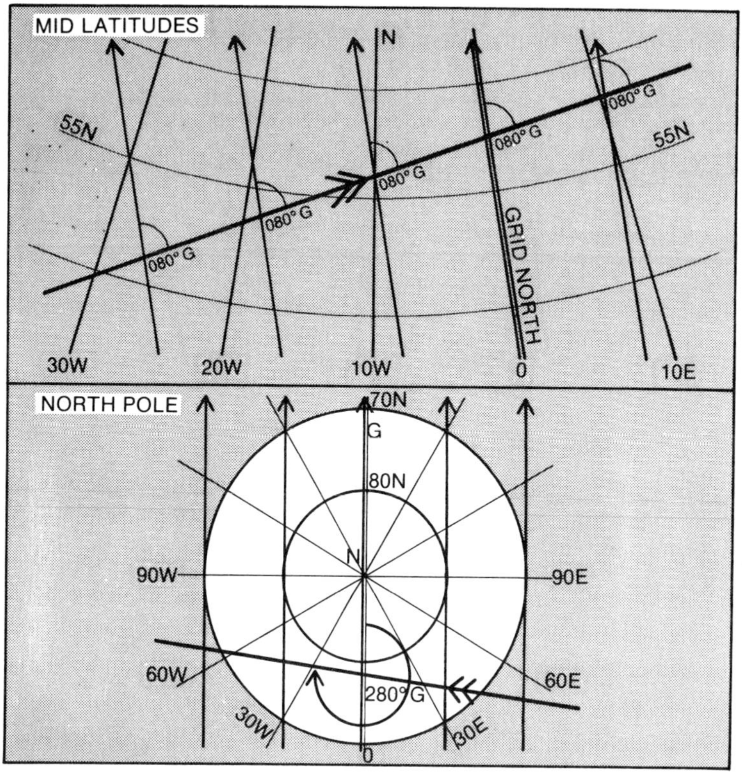

Fig. 5.5 Grid navigation.

Grid navigation in mid latitudes

On the North Atlantic, for example, Lambert's conformal chart, overprinted with the basic grid, was used to obtain the advantage of flying great circle tracks. Radio signals also travel along great circle paths and radio bearings could simply be plotted as straight lines on the chart. In this case the normal magnetic compass was used for steering track and it was necessary to convert the grid heading to a magnetic heading.

The angular difference between True North and Grid North at any point on the chart was known as grid convergence. The application of grid convergence converted grid direction to true direction, and variation converted true direction to magnetic direction. Grid convergence and variation were summed algebraically and applied as one correction known as grivation. Dotted lines known as isogrivs were printed on charts joining points of equal grivation. The grid heading required to maintain track was calculated, and the mean grivation along track taken from the chart and applied to give a constant indicated magnetic heading, which, when steered by the magnetic compass, maintained the aircraft as close to a great circle track as possible.

Grid navigation over the Poles

On Polar navigation the Polar stereographic chart was normally used. Like Lambert's chart, great circle tracks and radio bearings could be plotted as straight lines. Once again, the Greenwich Meridian was normally chosen as the Grid North datum. In Polar areas the magnetic compass is unreliable because of the proximity of the Magnetic Pole, and the compass was used in gyro mode. In this mode the gyro stabilized compass (free from magnetic influence) was arranged to maintain Compass North aligned with Grid North, the compass being suitably corrected for earth rotation. When a constant indicated grid heading was steered the aircraft maintained a great circle track.

Great care had to be taken when aligning the compass with Grid North at the aircraft's position, and when converting upper winds from true to grid direction and radio bearings from magnetic to grid. The pilots had no reference by which to steer the aircraft except the compass set by the navigator to the arbitrary datum of Grid North, and a simple error in alignment or heading could cause serious problems.

Fixing position

Position was normally fixed by obtaining three position lines for accuracy. A small triangle, known as a 'cocked hat', usually occurred, the position being considered to be in the middle of the triangle. It was seldom possible to obtain lines simultaneously, so a position line would normally have to be moved along the planned track at the aircraft's ground speed to the time of the required fix. It would take several minutes to obtain a fix, the time being noted against each position line as it was plotted and also against the final fix.

When within range of radio beacons, e.g. at the start and end of an ocean crossing, accurate fixes could be obtained from bearings and distance measuring equipment. Weather radar could also be tilted down to map the earth's surface, and a headland or island identified from a chart. Range and relative bearing from lines marked on the screen could then be noted and a fix obtained.

In mid ocean or desert, long range radio aids were required, and Loran (long range aid to navigation) was the most common source of position lines, particularly over the Atlantic and Pacific oceans. Loran determined position lines by noting the time difference between radio signals transmitted from distant master and slave stations. Hyperbolic position lines overprinted on charts were then used to plot position. Consol was also available over the oceans, and consisted of a pattern of dots and dashes transmitted from a distant station. Once the relevant Consol sector was established by 'dead' reckoning, counting of the dots or dashes would establish the position line on an overprinted Consol scale on the chart. On the North Atlantic, ocean station vessels (OSV) were also available, their positions on a grid being indicated by morse code. Radar bearings could be obtained from an operator on board once radio contact had been made.

Over most deserts, and sometimes in mid ocean, the aids described above were not available, and the navigator had to resort to astronomical navigation (Astro.) for fixing position by the stars. Unlike observations from a ship, however, the use of the horizon was unsatisfactory. Anyway, even if the horizon had been usable, the aircraft could hardly have waited until dusk or dawn when both stars and horizon are visible. Instead, the sextant was aligned with the vertical by positioning a bubble within the sextant between two parallel lines. The vertical obtained was, however, not quite true, because of displacement of the bubble by aircraft movement, and corrections had to be made.

Also, unlike at sea, the sky overhead was not visible, and the sextant had to be preset by calculations to find the required star. The flight navigator began by obtaining from tables the approximate altitude (e.g. the angle between the horizon and star) and azimuth (i.e. direction in relation to True North) of the star from the observer's position, before the star could be sighted and its precise altitude measured.

The sextant (Fig. 5.6) was placed in position through a double-door pressurized mounting situated on the flight deck roof. The first door was opened to allow the neck of the sextant to be inserted into the opening, and the second door then opened to ease the periscope tube out into the atmosphere. Because of the differential pressure a fairly strong force attempted to pull the sextant upwards and it had to be restrained from shooting up to the locked position. A flight engineer once unofficially designed an attachment for the end of a vacuum cleaning hose to fit into the sextant mounting. The idea was to open the double doors of the sextant mounting in the cruise and to use the hose like a vacuum cleaner to tidy the flight deck. On the first

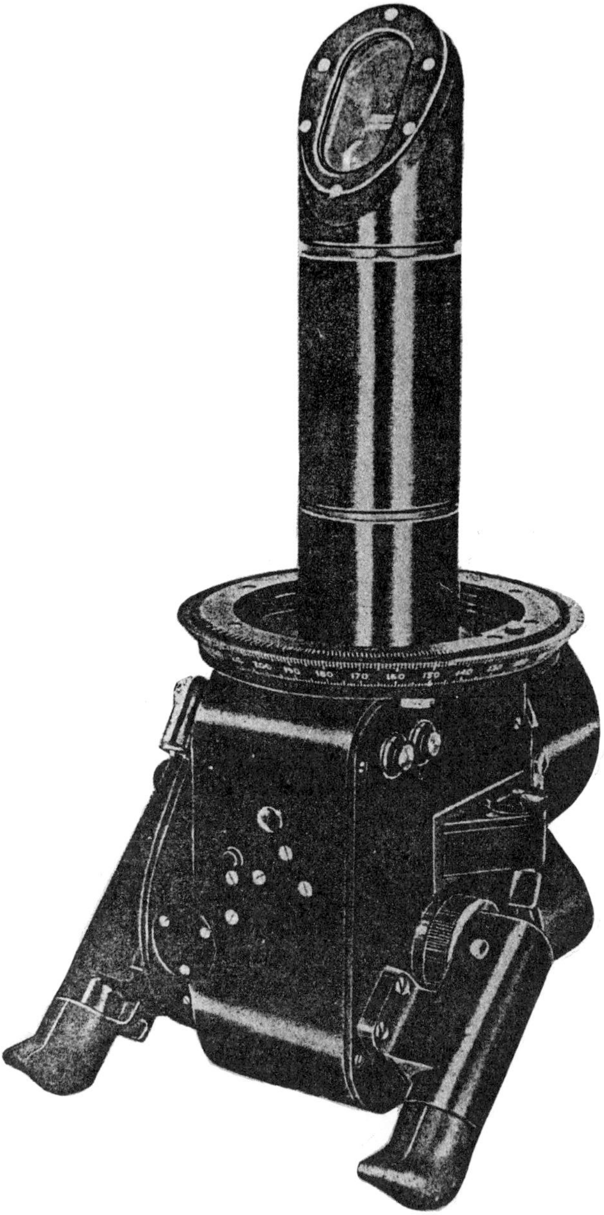

Fig. 5.6 The periscopic sextant.

trial the hose was attached to the sextant mounting and the double doors opened, but the suction was so strong it pulled the hose inside out and sucked out the complete unit!

The sextant was 'pointed' at the required star by setting the altitude calculated from the tables on the sextant, and by setting the sextant to the star azimuth by use of an azimuth ring on the neck of the sextant, which was aligned with true north. The star would then be in sight, but first had to be positively identified from the local star pattern within the sextant view before a 'shot' could be taken. Once identified, the star was 'shot' over a two-minute period to negate any inherent aircraft movement. For a star observed at 0200 GMT, the sighting commenced at 0159 and was completed by 0201. During this period the navigator would make slight adjustments to the sextant to maintain the star in view. At the end of the period the readings were summed mechanically within the sextant, and the average observed altitude reading obtained.

By comparing the calculated and observed altitudes a position line could be arrived at arithmetically and could then be plotted on the chart. Observations from another two stars allowed two more position lines to be plotted and a three-star fix of position could be obtained.

The workload for the navigator in navigating by the stars was very high. The aircraft moved at great speed and the navigator had quickly to navigate the aircraft back on to track to correct for any deviation, as well as complete the calculations for the astro. fix, keep a log going of the flight's progress, and 'shoot' the stars at the required time. Mistakes were easily made, so time was also required to check the calculations and to correct any errors. As a result it was only possible to obtain an astro. fix every 40 minutes, by which time the aircraft could well have travelled more than 350 n.m.

Although the majority of fixes were obtained from position lines of the same sort, i.e. three Loran or three stars, it was often necessary to use position lines from mixed sources when aids were scarce. In fact, in spite of the relatively sophisticated equipment available, it was not unusual for navigators to have to scratch around for position lines to obtain a fix, especially in the more remote areas where they were most needed. On occasions, two, or perhaps only one position line, were available. On the long daylight flights from Europe to the Caribbean, the sun was the only source of position line in the lower latitudes. The best the navigator could do was to use the sun line and the 'dead' reckoning position to establish a 'most probable position'. And this was only in the last decade, not the last century!

Route navigation

A useful route to illustrate the flight navigator's duties was from New York to London. Let's say, for example, the allocated track was from Gander, in Newfoundland, Canada, to Shannon, in Eire, routeing via 50N 50W, 52N 40W, 53N 30W, 53N 20W, 53N 15W. (There is nothing at these position reporting points — they are simply convenient co-ordinates on a chart). The pilots would first fly the aircraft from New York to Gander along the assigned airways. Under this form of pilot 'navigation' it was not necessary to plot the aircraft position precisely on a chart as position was easily monitored from the instruments. The pilots would simply steer the aircraft from radio beacon to radio beacon along the airways from New York to Gander, and also, at the other side of the 'pond', from Shannon to London, using airways charts as a driver would use a road map for guidance along a highway.

Approaching the Atlantic coast the navigator would mark the asssigned track on the navigational chart and would prepare to navigate the aircraft. The navigator's duties would commence overhead Gander. Meanwhile the pilots would simply fly the North Atlantic route track using the drift and ground speed information from the Doppler. The problem with the Doppler system, however, was that it was not sufficiently reliable for accurate navigation and needed to be monitored by the navigator. The position of the aircraft would be plotted on the chart by the navigator at regular intervals using a variety of navigational aids and would be compared with position as indicated by the Doppler. The navigator would then indicate to the pilots any corrections required to regain track. In this capacity the navigator monitored the progress of the aircraft as flown from Doppler information, rather than directed the navigation of the flight.

If the Doppler equipment failed, however, the pilots lost the reference by which to set heading and check the aircraft's progress. The pilots would then be blind to the navigation of the aircraft, and the navigator would have to resort to air plotting, the basic system of navigation mentioned in the previous section. The navigator would then come into his own, directing the navigation of the flight, and issuing headings and estimates whenever required.

A log of observations, estimates, position lines and fixes would be maintained throughout the flight. Within the first 200 n.m. the radio beacon at Gander, together with distance measuring equipment, would be available for a fix. Outside this distance the radio beacon would be out of range, and the navigator would make use of the Loran system (long range aid to navigation) available from stations on the North American continent and Greenland. At about 30W these Loran chains would be out of range, with the Loran chains in Europe not yet within range, and, with eastbound flights at night, a star fix would normally be obtained at this position. The European Loran chains would then be used until about 200 n.m. from Shannon, when the aircraft would once again be within range of the radio beacon and distance measuring equipment at Shannon. A last fix would be obtained and the aircraft navigated to overhead the beacon, at which point the navigator's duties would be complete. A typical crossing would involve about three and a half hours of navigating.

Pilot navigation

A number of radio aids are in use today for pilot guidance along airways and on approach to land, and are outlined below.

Non-directional beacon (NDB)

A non-directional beacon is the most basic of radio navigation aids, consisting simply of a radio transmitter emitting a signal on a published frequency, not unlike a broadcasting station. On board the aircraft the signal is received by a simple receiver modified for navigational purposes. The required frequency is selected on a digital display and the beacon identified by its morse code signals.

NDBs transmit in the low frequencey (LF) and medium frequency (MF) bands from about 200 kHz to 800 kHz. The range of the beacon is proportional to the square of the power (i.e. to double the range it is necessary to quadruple the power output) and varies from up to 500 n.m. over sea to 100 - 150 n.m. on land. NDBs are occasionally used to give centre line guidance on airways. More commonly, lower power NDBs, known as locators, with a range of about 10 n.m., are used as marker beacons on final approaches at airports.

Once the NDB has been tuned and identified it is necessary to establish from which direction the signal is being transmitted and this is achieved by equipment on the aircraft known as the automatic direction finder.

Automatic direction finder (ADF)

When the frequency of an NDB within range is selected the ADF simply aims a pointed needle in the direction of the beacon. The needle is superimposed on a compass rose to indicate magnetic bearings. Presented in this form the instrument is known as a radio magnetic indicator (RMI) and is shown in Fig. 5.7.

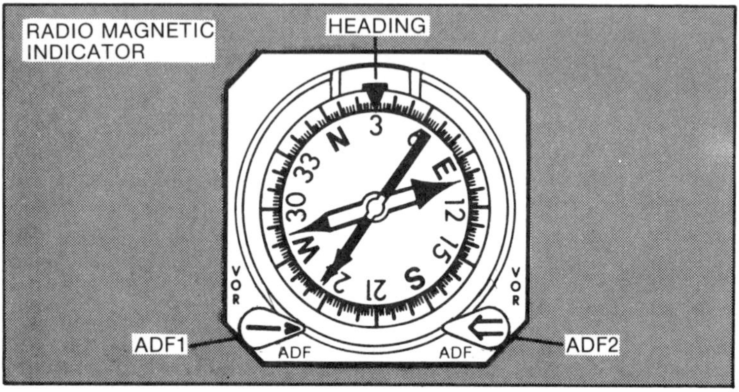

Fig. 5.7 Radio magnetic indicator.

The receiving equipment consists of a rotating loop with a single antenna placed within its influence. When the loop is positioned at right angles to a received radio wave the same phase is induced into both sides of the loop and no current flows. This position is known as the 'null point' and determines the line of the radio signal received. The single antenna resolves whether the incoming radio wave is from in front or behind the loop, thus establishing the direction of the beacon. Signals received are amplified to operate a small motor which automatically drives the system and the pointed needle to the 'null point' where signal strength is zero. The needle on the RMI then indicates the direction of the NDB from the aircraft.

Because of congestion in the LF and MF frequency bands beacons often transmit on the same frequency, and confusion is avoided by placing the beacons well apart with power outputs limited. At night, however, ionospheric activity increases the range and can cause interference between beacons on the same frequency. NDBs are also affected by atmospheric conditions such as thunderstorms, and can, on occasions, give quite erratic readings, the needles on the ADF momentarily pointing to clouds of high electrical activity instead of indicating the direction of the beacon.

Very high frequency omni-directional radio range (VOR)

The VOR is the most common type of radio beacon in use today.Its' main use is on airways, giving centre line guidance and indicating reporting points. As the name suggests VORs transmit in the very high frequency (VHF) band from 108.0 MHz to 117.9 MHz, the range being limited to line of sight, which is normally 200 nm.

The VOR transmits bearing information by the principle of phase comparison. The signal transmitted is frequency modulated (FM) at

Detling VOR, Kent, England.

30 Hz and is transmitted from an antenna system which rotates at 30 revolutions per second. Receivers on aircraft around the beacon therefore receive not only the FM signals at 30 Hz but also a signal which is amplitude modulated (AM) at 30 Hz because of the rotating antenna. The FM received has the same phase in all directions, but the phase of the AM received is different for every position of the aircraft from the beacon. The two modulations (FM and AM) are arranged to be in phase in the direction of Magnetic North from the beacon, and any phase difference between the modulations detected by the receiver therefore corresponds to the magnetic bearing of the aircraft from the beacon.

The VOR is tuned and identified in a similar manner to the NDB but displays bearing information in two ways; by a needle on a radio magnetic indicator (RMI) pointing towards the beacon, like the automatic direction finder RMI display, and also by a vertical orange

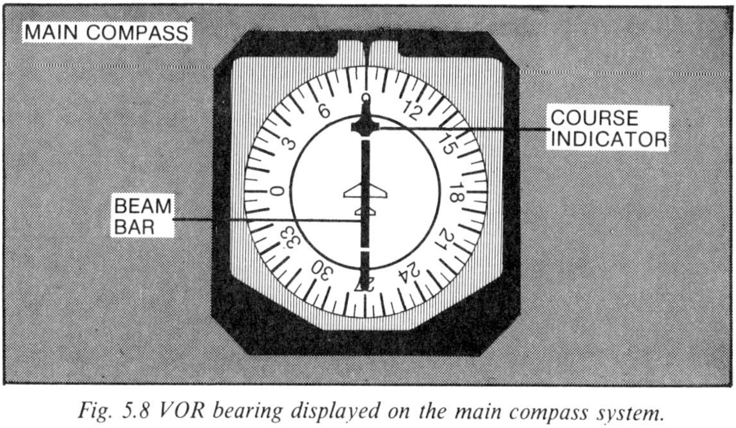

Fig. 5.8 VOR bearing displayed on the main compass system.

bar on the main compass system, known as a beam bar, which indicates the position of the required bearing from the VOR (Fig. 5.8). Radio beams transmitted by a VOR radiate outwards from the beacon like the spokes of a wheel, and are in fact called radials. The radio beam indicating magnetic east from the VOR, for example, is named the 090 degree radial (090°R). A radial from a VOR marking an airway centre line can be selected on a course indicator to display the centre of the airway. With the aircraft positioned on the airway centre line the vertical orange beam bar lies central in the main compass system. Such an instrument presentation displays to the pilots an instant picture of the position of an aircraft relative to a selected radial, and is a great improvement on the RMI where the needles simply point in the direction of the beacon.

VHF transmissions are not normally affected by atmospheric conditions, and when within range of a VOR information is of high accuracy. When outside the 200 n.m. range a striped red and white flag appears indicating that any information displayed is invalid.

Distance measuring equipment (DME)

DME is normally co-located with a VOR. It gives a highly accurate digital readout of the aircraft's distance from the beacon and is an invaluable aid to the pilot. DME operates in the ultra high frequency (UHF) band from 962 MHz to 1213 MHz, with a maximum range of about 300 nm.

DME is an example of secondary radar. Radio equipment on the aircraft, known as the airborne interrogator, sends out a stream of coded pulses of radio energy. When a pulse reaches a ground station, known as the ground transponder, it triggers off the transmitter which sends out a reply pulse to the receiver of the airborne interrogator. The time interval between transmission of the pulse and reception of the reply pulse is measured electronically, and the range of the beacon is automatically computed and displayed.

The range of frequencies from 962 MHz to 1213 MHz gives 252 frequencies, which are paired to provide 126 'channels'. Each channel consists of two frequencies spaced 63 MHz apart, one for air-to-ground interrogation and the other for ground-to-air response; e.g. Channel 1, air-to-ground, 1025 MHz; ground-to-air, 962 MHz. The use of different frequencies prevents the airborne interrogator accepting signals received from its own transmissions bounced back from the ground.

DME measures the slant difference from the aircraft to the beacon, which is slightly longer than the ground distance, but only about 0.5 n.m. more at 50 n.m. range. The maximum error occurs over the

Fig. 5.9 Localiser and glide slope antennae.

beacon when, instead of reading zero, the distance displays the height of the aircraft above the ground, e.g. 35,000 feet indicates 6.6 n.m.

The frequency of the DME is automatically tuned when the frequency of the co-located VOR is dialled, and the distance to go appears when the facility is within range.

Instrument landing system (ILS)

ILS consists of two separate radio signals giving both runway centre line guidance and descent profile guidance to the runway touchdown point. The runway centre line signal is known as the localiser (LOC), and the descent profile signal as the glideslope (G/S). (Glideslope is a misnomer here, as the aircraft is certainly not gliding. As can be heard on the approach to any airport, power is required to maintain the aircraft on the correct descent path with gear and flaps lowered.)

The localiser antenna is a large fence-like structure positioned at right angles to the runway at the far end of the approach. The glideslope antenna is positioned near the touch-down area offset to one side. (Fig. 5.9). Localiser frequencies lie in the VHF band from 108.1 to 111.9 MHz, and glideslope frequencies in the UHF band from 329.3 to 334.0 MHz. Each localiser frequency is paired with a glideslope frequency (e.g. LOC 108.5 MHz, G/S 335.0 MHz), and ILS frequencies are published in terms of localiser frequencies only. Selection of an ILS localiser frequency on the VHF radio navigation selector automatically tunes in the paired glideslope frequency. (The VHF selector is the same as that used for the VOR, and carries the full range of VOR/ILS localiser frequencies). ILS identification normally consists of a three- or four-letter group transmitted in morse code.

The localiser signal (Fig. 5.10) consists of two overlapping lobes of radio energy, transmitted horizontally on the same VHF frequency, which are differentiated by modulating the carrier waves at different frequencies. The precise centre line of the runway is defined where the two lobes of radio energy overlap. When the aircraft is to one side of the localiser centre line the received signal strength from the lobe of radio energy on that side is stronger than the other, and the aircraft position is accordingly indicated on the flight instrument display.

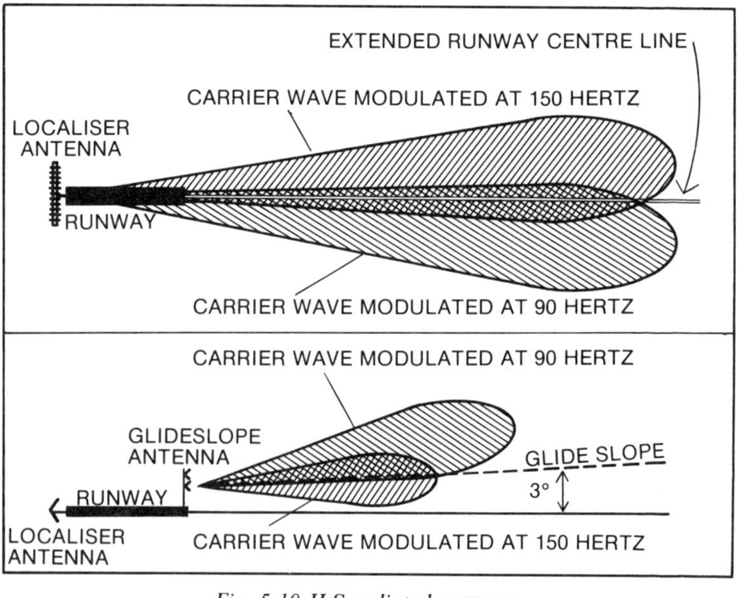

Fig. 5.10 ILS radiated patterns.

Similarly, the glideslope signal consists of two overlapping lobes of radio energy, transmitted vertically on the same UHF frequency, which are also differentiated by modulating the carrier waves at different frequencies. Where the lobes overlap defines the required descent profile for the final approach, which is sloped to the runway at about 3°. Above and below the 3° G/S the relevant lobe signal strength increases and thus also indicates on the flight instrument display the aircraft position high or low of the glideslope.

The ILS is calibrated for accurate landing guidance up to about 10 n.m., but it can in fact be picked up and used as an approach aid at 50 n.m. and more.

Marker beacons

Three marker beacons are positioned along the ILS to give an indication of range to touch-down. The beacons all transmit on 75 MHz and radiate a low power fan-shaped radio energy pattern, which can only be received directly above the transmitting antenna. An outer marker (OM) is placed at about 5 n.m. from the runway, a middle marker (MM) about 0.5 n.m., and an inner marker (IM) at the runway threshold. Tuning of the marker beacons is not necessary as they all transmit on the same frequency and each marker transmits its own particular identification. The outer marker signal is heard on the flight deck as a series of low-pitched dashes accompanied by a blue light flashing in unison; the middle marker as a series of alternating medium-pitched dots and dashes with an amber light flashing in unison; and the inner marker as a series of high-pitched dots with a white light flashing in unison. (The high approach speeds of modern jets has all but rendered the inner marker obsolete.)

The OM is frequently accompanied by a low powered non-directional beacon, known as a locator, positioned alongside (*see* NDB p.101). The combined beacon is known as a locator outer marker (LOM) and at many airports now is the only beacon remaining, being positioned about 3 - 6 n.m. from the runway (Fig. 5.11). It is, of course, necessary to select the frequency of the locator NDB, and the needles on the radio magnetic indicator (RMI) then point towards the beacon. As the aircraft passes overhead the LOM the low-pitched dashes are heard with the blue flashing light and the automatic direction finding (ADF) needles on the RMI swing round and point backwards. At this precise point a stop-watch can be started, and time to touch-down (already noted from the approach charts for the aircraft speed) can be commenced. As a precaution, a check of the ILS glideslope can also be made at this point by comparing aircraft height with the required height at the LOM noted from the approach chart.

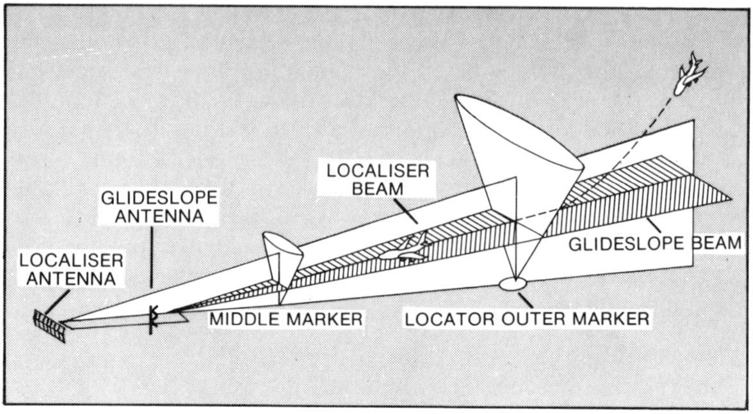

Fig. 5.11 The LOM.

Modern instrument landing systems at some major airports also include distance measuring equipment (DME). Selection of the ILS frequency automatically tunes in the associated DME, and the pilots have a continuous readout of distance to run to touch-down. Hence, when within range of such an ILS, at the selection of one frequency the pilots have displayed all the relevant information — runway centre line, descent profile guidance, and distance to touch-down — required for a safe and accurate approach.

Flight instrument display

Localiser and glideslope signals transmitted by the ILS ground equipment are received independently by antennae positioned on the fuselage of the aircraft. On the main compass system, the runway centre line position is indicated by a beam bar in a similar manner to the position of a radial when a VOR signal is being received. With the beam bar to the right of the instrument, the aircraft is left of the localiser and vice versa. When the beam bar is central the aircraft is positioned on the runway centre line (Fig. 5.12). The precise magnetic direction of the runway (e.g. at Chicago, O'Hare, the direction of the north easterly runway is 039°M) is set on a course indicator which positions the beam bar on the main compass to the required runway direction. The glideslope position is indicated by a yellow arrow on the right of the instrument. The flight director system also displays localiser position by a yellow vertical bar, and glideslope position by a yellow horizontal bar, on the artificial horizon. The yellow bars indicate in the same sense as the beam bar and yellow arrow. When the bars are crossed in the centre of the instrument the aircraft is positioned correctly on the ILS.

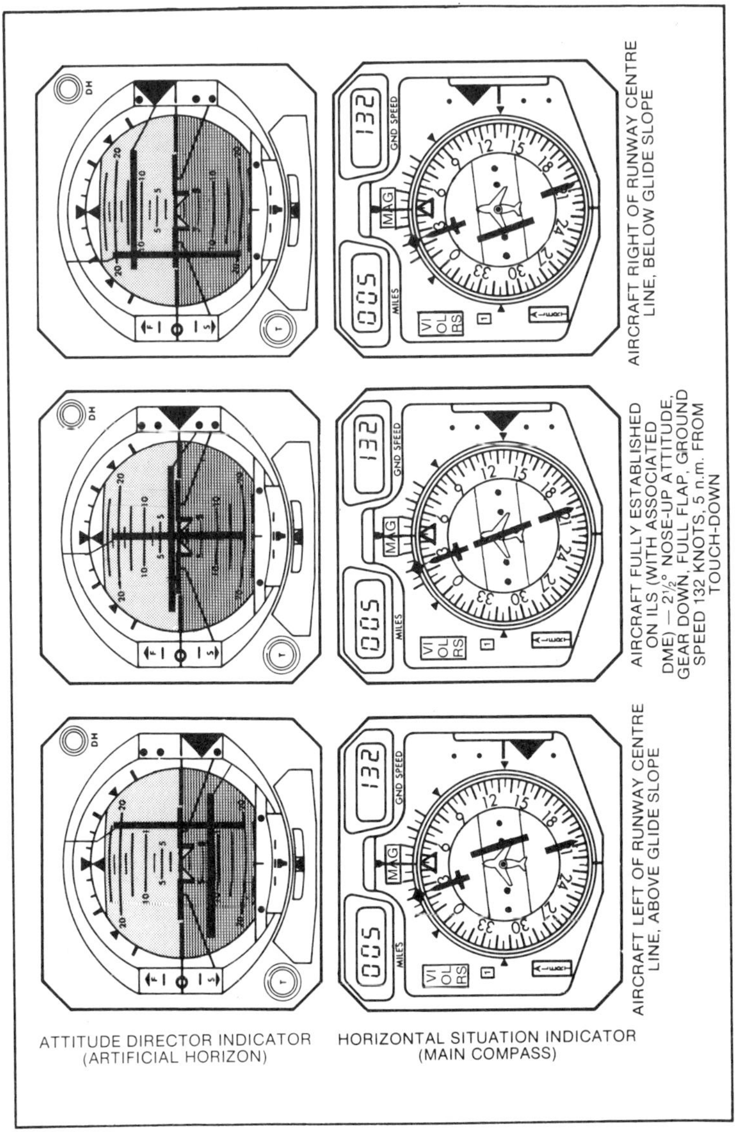

Fig. 5.12 ILS flight deck displays.

The ILS can be used for automatic approaches and automatic landings, with the autopilot engaged, but can also be flown by the pilot handling the aircraft and simply responding to the indications on the instruments to maintain the aircraft on the ILS. Hand-flying the ILS precisely is a skilful operation, but most pilots prefer hand-flying the approach, when operating conditions allow, to simplify procedures and maintain handling practice. Automatic approaches tend to be long-winded as the aircraft requires time to settle in each phase of the approach. Also, flying the aircraft by the autopilot switching can be a complicated procedure, especially if failures occur and warning lights begin to flash. However, if the cloud base is low, e.g. 200 feet, then automatic approaches become mandatory as the autopilot flies the ILS more accurately. Air traffic control then positions aircraft further out, by radar, before commencing the approach. Automatic landings are designed for foggy conditions, when winds are light, and cannot be attempted when cross-winds are above 15 knots. At such times the aircraft is required to be hand-flown.

Computer navigation

Inertial navigation system (INS)

Inertial navigation system is a self-contained airborne unit which senses aircraft movement by means of accelerometers placed on gyro-stablilized platforms and which continuously computes and displays navigational data. Signals from the INS also supply a stable reference datum for certain flight instruments and for autopilot function. Normally three separate systems are installed on each aircraft.

The heart of the INS is the inertial reference unit, which houses the accelerometers on gyro-stabilized platforms. An accelerometer is basically a small sprung mass whose movement is sensed electronically during periods of acceleration or deceleration, the signal produced being passed to a computer for processing.

A gyro is a spinning device which has the property of remaining in a fixed position in space. One example, where a small wheel is fitted horizontally within a light frame, is used as a toy. The wheel is spun by wrapping around the spindle a piece of twine and giving it a sharp pull. With the wheel spinning, the gyro can be placed at any angle, and the axis remains in a fixed position. If placed vertically on the back of the hand, the hand can be rocked from side to side, like the movement of an aircraft, and the gyro remains stationary (Fig. 5.13). In the artificial horizon, for example (*see* Flight Instruments, p.119), the horizontal position of the spinning gyro wheel is the reference for

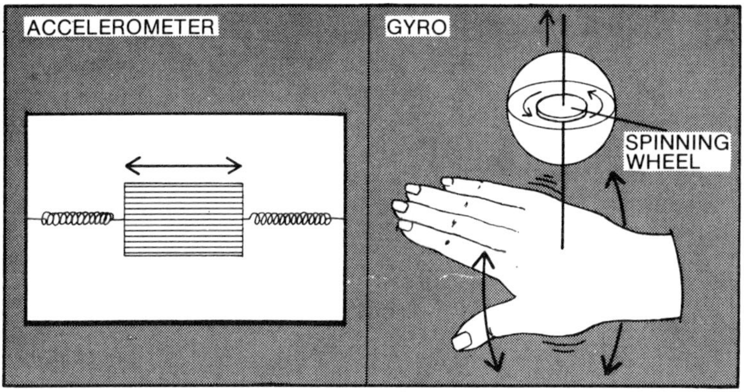

Fig. 5.13 The action of a gyro.

the horizon on the instrument, and remains fixed in line with the actual horizon as the aircraft rolls and pitches. The gyros in the INS employ the same basic principle as the gyro toy described above, but in this case the central stabilizing gyro wheel is spun electronically at around 100,000 rpm.

The gyro property of rigidity also creates a problem. The stabilized gyro indicates the local horizon at a fixed point in space, but as the earth rotates, or as the aircraft moves over the surface of the earth, the local horizon reference changes and the gyro remains fixed, indicating the horizon at the original point. To overcome the problem the gyro-stabilized platforms of the INS are electronically controlled by means of torque motors to maintain the local horizontal plane, thereby utilizing the property of rigidity while continuously adjusting the gyro to the local horizontal reference. Stable conditions for the accelerometers and a local horizontal reference for the relevant flight instruments are thus maintained. (The principle is similar to that employed during INS alignment. See next section.)

Two sets of accelerometer/gyro pairs sensing movement on the horizontal plane are mounted at right angles to each other on a horizontal gyro-stabilized platform. The platform is rotated at one revolution per minute to reduce any error effect of accelerometer/gyro misalignment. Distance travelled through the air in relation to that over the ground increases with height, so a single accelerometer/gyro pair is also mounted in the vertical plane to measure vertical movement of the aircraft. Stability of the system is achieved by mounting the complete platform group on a gimballed assembly.

In the navigation mode, any aircraft movement in the horizontal

INS Control/Display Unit.

or vertical planes is sensed by the accelerometers, which produce signals proportional to the accelerations detected. These output signals are then passed to a computer which determines the required data (such as speed, distance, track, etc.) for the navigation of the aircraft. The computer also passes signals to the torque motors positioned on the gimballed assembly, which operate to maintain the gyro platform level with the local horizon.

Change in attitude of the aircraft in pitch, roll and yaw is also detected by small sensing units, known as synchros, strategically placed on the gimballed assembly. The synchros produce signals proportional to change of aircraft attitude and are fed via a computer to the relevant flight instruments for attitude display, and to the autopilot for aircraft guidance.

INS alignment

All the information on the shape of the earth, earth movement and direction of True North, etc., is held in the computer program. The position of the aircraft (on its arrival from the last flight) is stored in the computer memory, so the INS has all the information required to navigate the aircraft, even before it is switched on. Small errors, however, do creep into the operation, and after a long flight the INS may 'think' that it is perhaps 5 or 10 n.m. away from the actual position of the aircraft. This problem is resolved during an alignment period when the actual position of the aircraft is inserted into the INS. One of the flight crew inserts the precise present position of the aircraft at the terminal building (taken from the airport charts; e.g. New York, Kennedy, 40° 38.9′, W 073° 46.9′) into the computer by use of a numbered keyboard on a control/display unit. The INS then compares the actual position inserted by the pilot with where it thinks it is, and makes the necessary corrections in an attempt to eliminate the error on the next flight. It also works the other way. If the pilot inserts the wrong position, the computer compares this with the memory position, and rejects the inserted position by illuminating a red warning light on the display unit if the error is significant.

The datum for all INS navigational measurement is True North, and the computer has to recalculate the direction of True North from the actual position of the aircraft to complete the necessary re-alignment. The computer recalculates the direction of True North in a very basic but clever way.

Let's imagine, for simplification, than an aircraft is positioned on

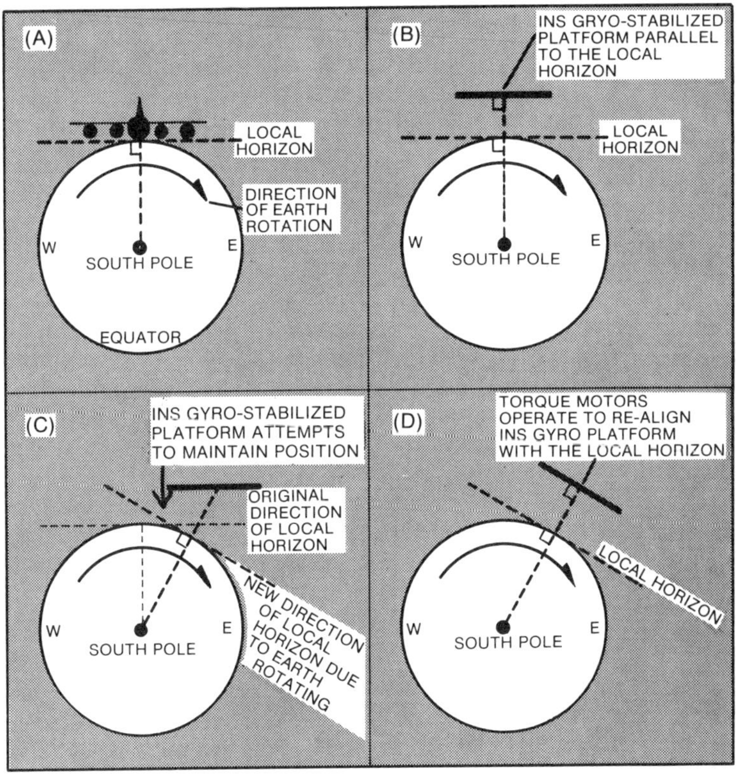

Fig. 5.14 INS alignment.

the Equator, heading north, with the earth being looked at from below the South Pole (Fig. 5.14a). The calculation can, of course, be made from almost anywhere on earth, except near the Poles, and for any aircraft heading. The gyro-stabilized platform is positioned parallel to the local horizontal as shown in Fig. 5.14b. As the earth rotates, the gyros of the gyro-stabilized platform hold the platform fixed in space and attempt to maintain the platform parallel to the direction of the original horizon (Fig. 5.14c). The accelerometers on the platform, however, sense a component of movement with earth motion. The proportional acceleration signals produced are processed by the computer, and the relevant signals then relayed to the torque motors on the gimballed assembly, which drive the platform level to the new local horizontal (Fig. 5.14d).

During the alignment process, as the gyro platform attempts to dip to the west, and is maintained parallel to the new local horizon level, and attempts to dip to the west again, and so on, the INS computes the direction of True North to be at right angles to the direction of attempted dip (i.e. west), and so establishes the direction of True North. Therefore, in a period of approximately 13 minutes, simply by the movement of the earth, the computer recalculates the calibration values required to maintain the platform level to the local horizon, establishes the direction of True North and determines any instrument error.

INS route navigation

A facility on the autopilot allows the INS automatically to navigate the flight, but first the pilots have to tell the computer the aircraft routeing by keying into the INS the exact latitude and longitude of reporting points or radio beacons along the way. Such INS positions are known as way-points, and up to nine can be inserted at any one time, the sequence being repeated as the aircraft progresses. When the autopilot is engaged and INS selected, the autopilot automatically steers the aircraft, with continuous slight changes in heading, on a great circle track direct to the first way-point position, then automatically turns the aircraft for the next, and so on. The navigation function of the pilots, therefore, is to monitor the performance of the INS as it navigates the aircraft via the autopilot.

The navigation of the aircraft by INS is usually commenced by about 10,000 feet when the autopilot is engaged and the INS mode selected, but it is also frequently used to obtain navigational data, even when the aircraft is being hand-flown.

On airways, the INS is used to navigate the aircraft via the autopilot, but its performance is monitored by tuning radio beacons and by selecting courses along the route as normal. On long range flights, across the oceans and deserts, the INS comes into its own, navigating, quite independently, with great accuracy. On all flights the information inserted into the INS is checked carefully by each crew member in turn, and is checked again at each way-point as the flight progresses. When within the terminal area of an airport, or within a congested airways area, especially after a long flight, the position of the aircraft as indicated by the INS may be in error by a few miles, and is not sufficiently accurate for navigation. (Some systems can be automatically corrected, i.e. updated, by use of distance measuring equipment.) At this stage, if update facilities are not available, the INS mode is normally disengaged, and the aircraft navigated from beacon to beacon in the conventional manner, using the heading mode of the autopilot. When in the approach phase of the flight, especially when hand-flying the aircraft on the final approach, the INS is used to obtain such information as wind, drift, and ground speed.

Omega

Omega is a precise modern long range navigation aid using ground-based transmitters. Eight stations cover the world, transmitting on 10.2 kHz in the very low frequency (VLF) band. A sequence of three one-second pulses, each on a different frequency, is transmitted by every station within a ten-second period. The signals are synchronized at each station by the use of extremely accurate atomic clocks. On board the aircraft a small computer stored with information on the positions of the various beacons processes the signals received and calculates the aircraft's precise position. Although, once again, relying on ground-based equipment, the Omega system navigates world-wide to an accuracy of 2 n.m., which is an improvement on present INS capability, and the use of Omega as a navigational aid is increasing on non-INS equipped aircraft.

Laser gyroscope

The latest development in navigation is the laser gyroscope, another self-contained navigation aid independent of ground-based or satellite transmitters. Two contra-rotating laser beams are transmitted round a triangular tube with angled mirrors, as shown in Fig. 5.15. During aircraft movement angular rotation rate is measured by sensing frequency shift in the beams, thus determining angular position. Three such laser gyroscopes are arranged at right angles to each other to determine aircraft attitude in pitch, roll and yaw. Three accelerometers similarly mounted detect accelerations along the same axis. Attitude and acceleration data are resolved to provide position information that is fed to a flight management computer for navigation. Navigation performance is similar to the INS, the main advantages being digital output ability (desirable in modern avionics) and lack of moving parts. Equipment costs and power consumption are also lower in comparison to INS.

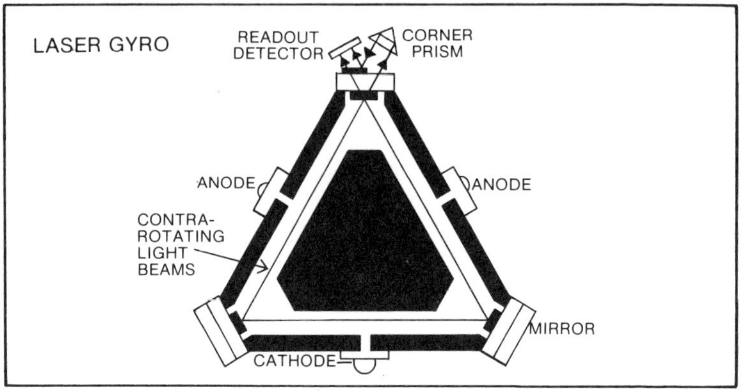

Fig. 5.15 The laser gyroscope.

Chapter 6
Flight Instruments

Four main flight instruments form a standard group known as the basic 'T' pattern, which consists of the artificial horizon (AH), the compass (C), the airspeed indicator (ASI), and the altimeter (ALT). Additional instruments such as the radio altimeter (RA), vertical speed indicator (VSI), turn and slip indicator (T & S), and radio magnetic indicator (RMI — *see* Navigation p.101) are also required, but the basic 'T' pattern remains (Fig. 6.1). The artificial horizon and the compass are gyro-stabilized instruments that receive stable reference signals from the inertial navigation system (INS) (*see* Navigation p.110). (On non-INS equipped aircraft each instrument contains its own stabilizing gyro). The airspeed indicator, altimeter, and vertical speed indicator are air pressure instruments. The radio altimeter operates by bouncing radio signals from the ground.

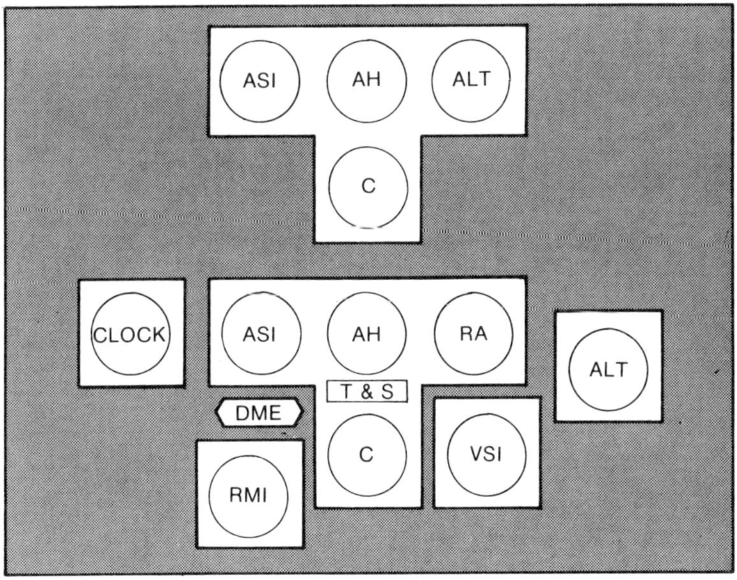

Fig. 6.1 Flying instruments: the 'T' pattern.

The artificial horizon is grouped with the turn and slip indicator. Flight director (FD) bars can also be selected to display flight profile requirements. In such form the instrument group is known collectively as the attitude director indicator (ADI). The main compass system not only indicates heading but also displays aircraft drift, VOR radial position, and 'raw' ILS localiser and glide path information. In such form the instrument group is known collectively as the horizontal situation indicator (HSI).

Attitude director indicator (ADI)
All big jets are flown with reference to flight instruments, the focal point being the ADI (Fig. 6.2). At night or in cloud the requirement for a horizontal reference is obvious, but the instrument is also used in clear conditions, even when a horizon is sharply defined. On large jet aircraft the wings are well aft of the pilot and out of view of the flight deck, and maintaining level flight visually is difficult.

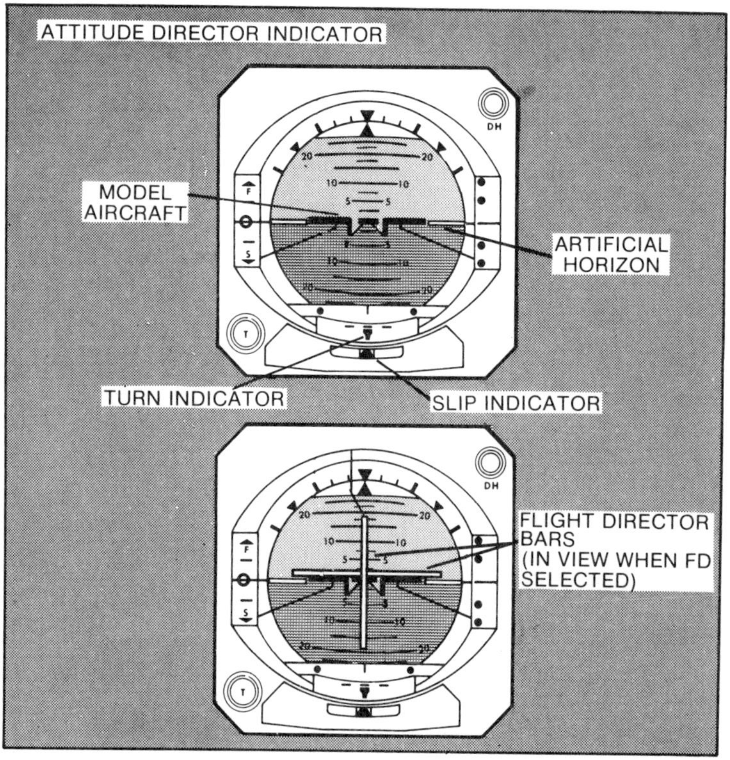

Fig. 6.2 Attitude director indicator.

The basic 'T' panel with Attitude Director Indicator centre and the Horizontal Situation Indicator below.

The orange model aircraft remains fixed to the aircraft structure and moves in sympathy, while the artificial horizon maintains the local horizontal. There is one ADI on each pilot's panel and a standby artificial horizon on the centre panel, by the Captain. The number of artificial horizons, i.e. three, is significant, as anyone with two watches will know. If one is wrong it's not known which is correct. With three watches it can be reasonably assumed that any two indicating the same time are right. Likewise, if any one artificial horizon malfunctions, the failure can be observed by comparison with the other two instruments.

The **turn and slip indictor (T & S)** forms the lower section of the ADI. The turn indicator is used to indicate the **rate** of turn (i.e. number of degrees of turn per minute) as, for example, when flying a race track pattern over a holding beacon. A rate one turn is defined as one in which a turn of 180° is completed in one minute, and a rate two as one in which a turn of 360° (a circle) is completed in one minute.

The slip indicator is useful in maintaining balanced flight. The instrument is not unlike a curved spirit level, and consists of a small white ball placed within a fluid filled glass tube. When the ball is central between two markers the various forces experienced in flight are in equilibrium and the aircraft is in balanced flight. During turns, if insufficient bank angle is applied for a particular speed, the forces experienced deflect the ball away from the direction of turn, the flight is unbalanced and the aircraft is said to be skidding. If too much bank angle is applied the forces experienced deflect the ball towards the direction of turn and the aircraft is said to be slipping (Fig. 6.3). The instrument is also useful during asymmetric flight (i.e. with engine failure) when the yaw experienced deflects the ball from the central position. The application of rudder counteracts the power imbalance and straightens the aircraft. The ball returns to the central position, and balanced flight is achieved.

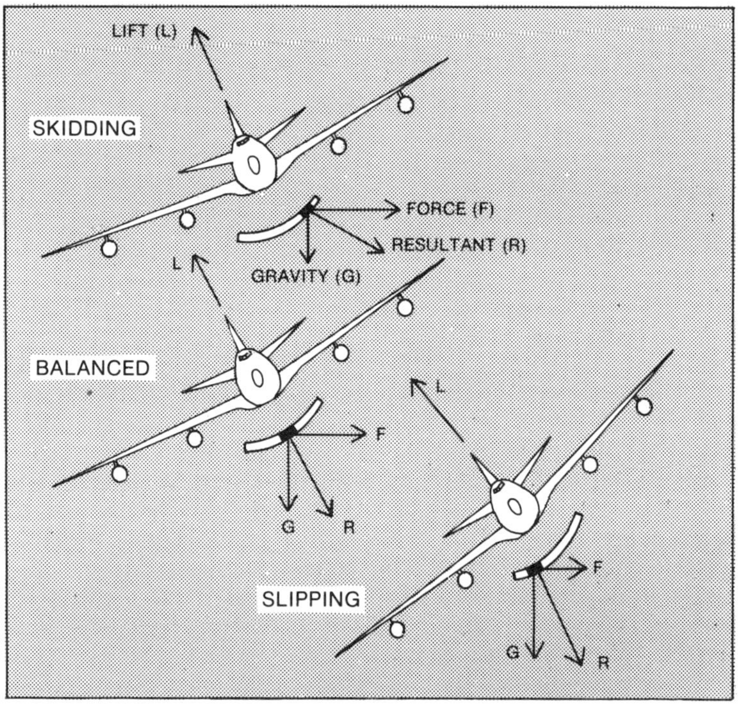

Fig. 6.3 Effects of different bank angles on the turn and slip indicator.

The **flight director (FD)** can be switched on to display two yellow bars — one vertical, the other horizontal — on the ADI. Its function is to display to the pilot the required flight profile and to 'direct' the pilot along the desired flight path when hand flying the aircraft. The yellow FD bars receive information from the autopilot computer via the autopilot mode selectors. The vertical yellow bar can indicate heading, a selected VOR radial, an ILS localiser course, or a programmed INS track. To maintain the required flight path the pilot simply flies the aircraft to hold the bar central in the instrument. The horizontal yellow bar can indicate the pitch attitude required to maintain height, speed, vertical speed, or ILS glide slope. By positioning the model aircraft below the horizontal yellow bar, the pilot can hold the required pitch attitude for the selected flight profile.

Horizontal situation indicator (HSI)

A basic compass consists of a Magnetic North-seeking compass rose pivoted within a damping fluid. This instrument is subject to acceleration and turning errors, and oscillates wildly during turns. (Any rally driver knows the problems of attempting to read an automobile compass on the move.) In the higher latitudes it's also affected by the earth's magnetic field which dips the compass rose towards the Pole. The basic compass is therefore unsuitable for heading reference and a gyro-stabilized compass is required. A remote magnetic sensor, usually positioned in a wing tip, supplies the gyro compass with magnetic information and when operating in this mode the compass system is said to be 'slaved' to Magnetic North. The gyro compass can also be selected to INS to indicate True North. There are three compasses on the flight deck, one horizontal situation indicator compass system on each pilot's panel, and a standby on the central window post, which is a basic compass subject to the errors outlined above.

The radio magnetic indicator (RMI — *see* Navigation p. 101) compass rose is permanently aligned with Magnetic North. On Polar flights from Europe to Alaska, the main compass system is selected to True North, while the RMI compass, although unreliable, continues to attempt to indicate Magnetic North. The track passes between True and Magnetic North, and the incongruous situation arises where the main compass points in one direction to True North and the RMI compass in exactly the opposite direction, to Magnetic North.

Superimposed on the HSI compass system (Fig. 6.4) is an orange course indicator and beam bar. With a VOR selected the beam bar displays radial information (*see* Navigation p. 103) and with ILS selected the beam bar displays localiser position. A yellow arrow on the right of the instrument also indicates glide slope (*see* Navigation p. 109).

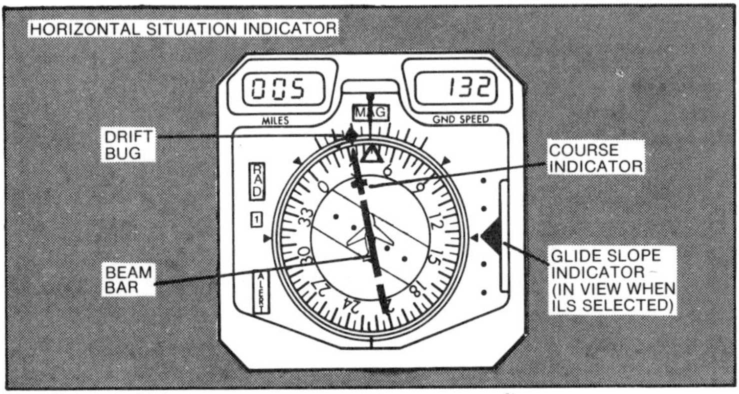

Fig. 6.4 Horizontal situation indicator.

The airspeed indicator (ASI)

An aircraft in motion is subject to a pressure force from the air known as dynamic pressure, and to the influence of static pressure. The dynamic pressure experienced is an indication of airspeed and is measured by a device known as a Pitot tube, named after its inventor Henri Pitot, an eighteenth century French hydraulic engineer. The tube is usually positioned near the nose of the aircraft in an area of relatively stable airflow, and points in the direction of travel. It is also heated to prevent icing. Static pressure measured at a static vent on the side of the Pitot tube is fed to a sealed ASI instrument case, and static and dynamic pressures measured at the open end of the Pitot tube are fed to a sensitive capsule situated within the case (Fig. 6.5). Since static pressure is evident within both the case and the capsule, static pressure is cancelled out and only dynamic pressure is communicated via the linkage to the instrument.

The indicated airspeed (IAS) obtained and displayed in knots (nautical miles per hour) is subject to a number of errors. At high altitude the air is very thin and of low density, which results in the airspeed under-reading. Also, the compressibility of air (ie. a volume of air can be reduced with a resultant increase in pressure) results in air entering the Pitot tube (at speeds above 300 knots) becoming compressed, causing the ASI to over-read. Corrections have to be applied to the indicated airspeed reading to compensate for density and compressibility errors and to obtain the true airspeed (TAS) of the aircraft through the air. At normal jet cruising altitudes compressibility error is negligible, but density error is such that the ASI under-reads by a significant amount and is unsuitable for airspeed indications (e.g. at 35,000 feet with an outside air

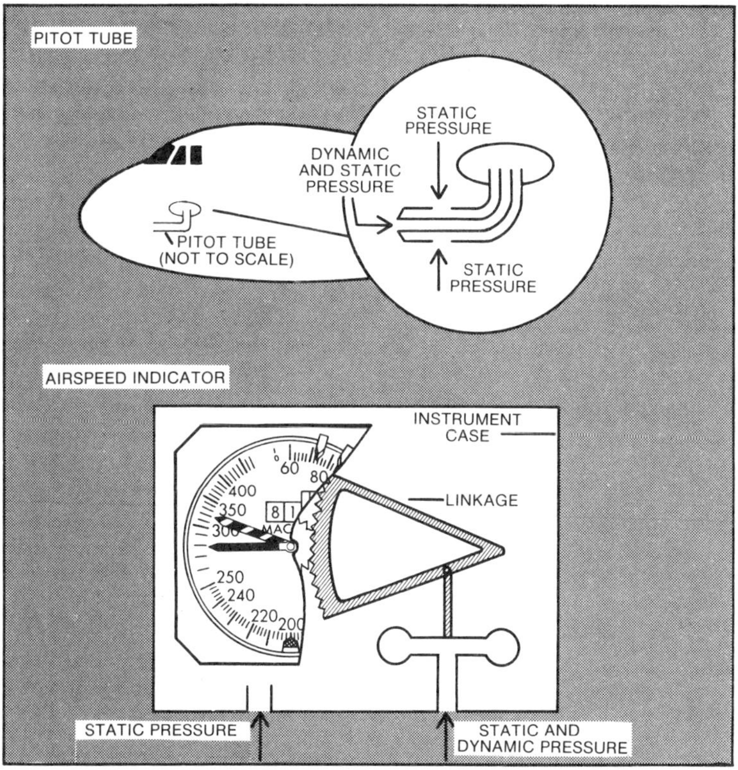

Fig. 6.5 The Pitot tube and the airspeed indicator (simplified).

Pitot Tubes.

temperature (OAT) of -60° Centigrade the indicated airspeed is 280 knots while the true airspeed is 480 knots). For cruising speeds above about 25,000 feet the Mach meter is used (see next section).

The aircraft operates within a specific airspeed envelope. If the aircraft flies too fast the airflow over the top surface of the wings reaches a point where its speed becomes supersonic, producing shock waves that result in a condition known as 'high speed buffet', the effects of which can be felt and which can damage the aircraft. The speed at which high speed buffet occurs varies with altitude, and is indicated on the ASI by a red and white striped needle known as the 'barber's pole', which is effectively the maximum speed at which the aircraft can fly for that particular height (Fig. 6.6). If the aircraft exceeds the maximum speed a warning is heard in the form of a clacking sound on the flight deck. Speed control is normally monitored by the flight engineer, and if the clacker is allowed to sound tradition demands that he stands the first round in the bar!

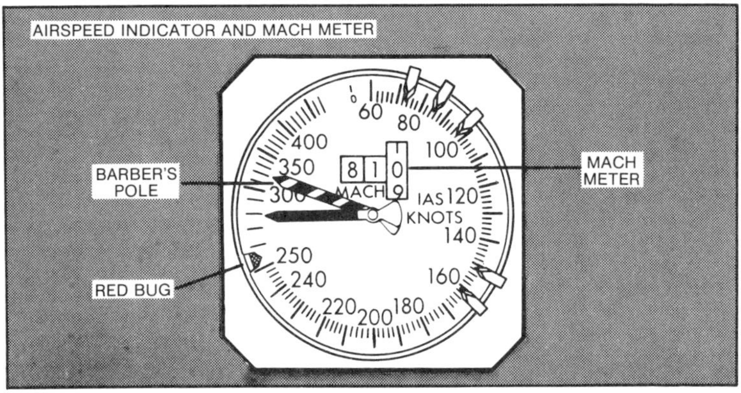

Fig. 6.6 Airspeed indicator and Mach meter.

At the opposite end of the scale, if the aircraft flies too slowly, the smooth air flow over the top surface of the wings starts to break away and to become turbulent. The aircraft shakes and judders, and a condition known as 'low speed buffet' is experienced. The speed at which low speed buffet occurs in terms of indicated airspeed is extracted from graphs for the particular height and weight of the aircraft, and is marked on the ASI by a red bug. This low speed buffet value is frequently updated by the crew as aircraft weight reduces with fuel consumption, or if the aircraft climbs or descends to a different level.

All cruise speeds for a particular height and weight lie within the range of speeds encompassed by the low speed buffet as marked by the red bug, and the high speed buffet as indicated by the barber's pole on the ASI. As the height increases the aircraft has to fly faster to maintain the same lift from the thinner air and to stay above the low speed buffet value. This higher speed brings the aircraft closer to the high speed buffet value for that particular height, and eventually a point is reached at which high speed buffet is experienced if the aircraft flies a few knots faster and low speed buffet is experienced if the aircraft flies a few knots slower. The point at which this occurs, known in crew jargon as "coffin corner", is the theoretical maximum operating height of the aircraft and for the Boeing 747 is about 45,000 feet. However, at normal operating weights the Boeing 747 is rarely flown above 39,000 feet. (Engine maximum operating height is around 42,000 feet.)

The Mach meter

Because of the density error effect mentioned earlier the airspeed indicator is not suitable for speed indication above about 25,000 feet. Instead the speed of an aircraft is given as the ratio of the true airspeed (TAS) of the aircraft to the local speed of sound, and is known as the Mach number, after the Austrian physicist Ernst Mach. The speed of sound is not constant but decreases with drop in temperature, which normally occurs with increase in altitude, being 661 knots at sea level in the standard atmosphere and 589 knots at 30,000 feet. The speed of sound is given the value 1.00 at any height, and therefore an indicated Mach number of, for example, 0.84 at 35,000 feet, indicates that the aircraft is travelling at 84 per cent of the value of the speed of sound at that height. Cruise Mach numbers for large jets at levels from 28,000 feet to 39,000 feet range from about 0.80 to 0.85 Mach. The maximum for the Boeing 747 is 0.90 Mach.

The altimeter

In a simple altimeter, static pressure is fed to an altimeter instrument case. As the aircraft climbs the air pressure reduces, and a partially evacuated capsule within the instrument expands. This expansion is transmitted through a mechanical linkage to a pointer that indicates the aircraft's height. The altimeter is calibrated for the standard atmosphere, which in practice is seldom experienced, and on modern aircraft corrections are applied to the altimeter by computer to improve accuracy. (Fig. 6.7).

Accurate height measurement by aircraft has always been a problem. The radio altimeter is a very precise piece of equipment but

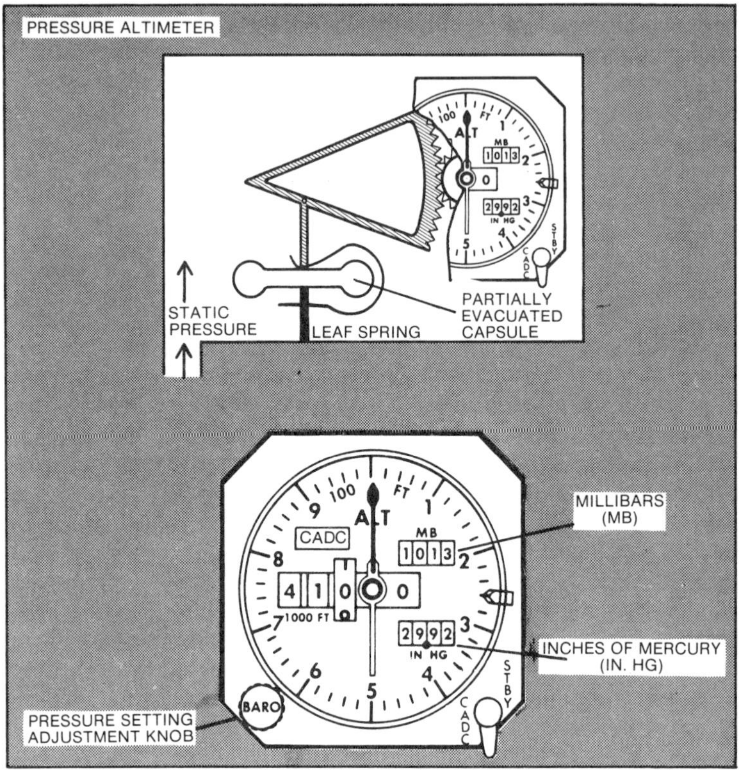

Fig. 6.7 Pressure altimeter (simplified diagram).

is not suitable for cruising in level flight. If the same height was maintained above the ground by the instrument, the aircraft would follow the contours of the ground and would climb and descend over mountains and valleys. Radio altimeters are only used on the final approach for accurate height measurement above the runway threshold, and in fact are only activated below 2500 feet.

The pressure altimeter is the primary instrument for height indication, but since it measures air pressure, which varies throughout the day in one place and also varies from area to area, further problems are encountered. When flying within a particular region, the important height indication required is the altitude of the aircraft above mean sea level (MSL). Since all charts indicate high ground as a height above MSL the aircraft can fly by the altimeter at an altitude which maintains the aircraft clear of high ground. To indicate the correct altitude in feet requires the setting of the current area MSL pressure on a small sub scale on the altimeter by means of an adjustment knob. This current MSL pressure setting is passed to the aircraft by the relevant air traffic control centre. In North America pressure settings are given in inches of mercury (in. Hg) and in most of the rest of the world in millibars (mb), requiring two subscales on the altimeter to accommodate both.

The MSL pressure setting is known simply in the USA as the altimeter setting and elsewhere as the QNH, which is a throwback to the old radio 'Q' code, and doesn't actually mean anything. When landing at an airport with QNH set, the altimeter still indicates altitude above MSL. It can be an advantage to set the altimeter to read the height of the aircraft above the airport and, therefore, to indicate zero on landing. To achieve this an airport elevation pressure setting, known as the QFE, is passed to the aircraft by airport tower (or is calculated from tables) and is set on the altimeters by the pilots at some time on the approach. QFE is not used on the Boeing 747, height measurement on final approach being read from radio altimeters.

In the cruise local pressure settings are not required as above a certain height a standard pressure setting of 1013.2 millibars or 29.92 inches of mercury is set on all altimeters. This standard setting represents the value of MSL pressure on an average day. When cruising at 35,000 feet, for example, with standard set, the aircraft will not be precisely at this height (unless the standard setting happens to be the MSL pressure in the area at that time), but with all altimeters adjusted to the standard setting, separation between aircraft is accurately maintained. The altitude at which the altimeter is changed from the local altimeter setting to the standard setting is known as the transition altitude (TA) and varies throughout the world. The situation is very confused, with every country, and in some countries each airport, having its own idea of what transition altitude should be. The height varies from anything between 2000 feet to 18,000 feet. Up to TA, altitudes are given in thousands of feet, and thereafter in flight levels, where the last two zeros are omitted. e.g. 35,000 feet becomes flight level 350 (FL 350). On the descent the level at which the altimeter is changed from the standard setting to the altimeter setting (QNH) is known as the transition level (TL) and varies according to the TA.

The correct use of the terms, elevation, height, altitude and flight level are shown in Fig. 6.8, although they are often used loosely. Standard cruise flight levels for the big jets from FL 280 are FLs 290, 330, 370 and 410 eastbound, and 280, 310, 350 and 390 westbound, which for most levels allows 4000 feet separation between aircraft travelling in the same direction, and 2000 feet between aircraft travelling in the opposite direction. Exceptions are on the North

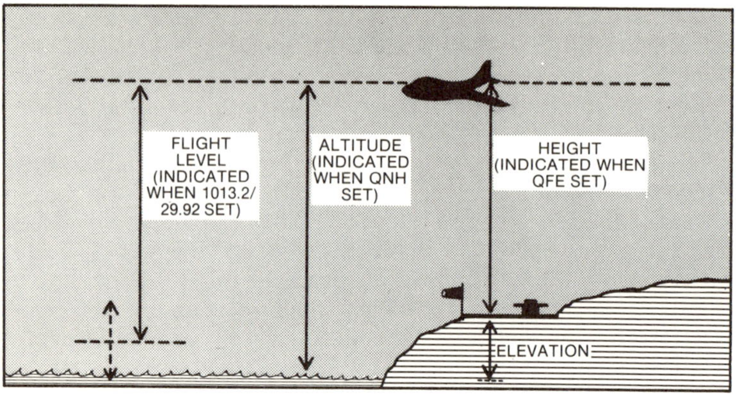

Fig. 6.8 Elevation, height, altitude and flight level.

Atlantic track system and, for example, in the more remote areas of Australia, where all the flights are travelling in the same direction at the same time, and non-standard flight levels of 300, 320, 340, 360, and 380 are frequently allocated, as well as opposite direction levels. In the USSR and China aircraft fly at metre flight levels. For example, in Russia an eastbound flight level of 9000 metres requires a western-equipped aircraft to fly at 29,550 feet. Although China and Russia share a common border, co-operation between the two countries is minimal and, believe it or not, some Russian and Chinese flight levels are actually the same for aircraft travelling in different directions on the same airway, e.g. FL 6000 metres for Russian westbound flights and Chinese eastbound flights, and flight level 13,000 metres for Russian eastbound flights and Chinese westbound flights. It is just as well the border is closed!

Vertical speed indicator

In a simple vertical speed indicator (VSI) static pressure is fed to a capsule, and also to the instrument case via a restrictive choke (Fig. 6.9). As the aircraft climbs or descends the static pressure in the capsule changes more quickly than the static pressure in the instrument case because of the restrictive choke, and the difference between the two is calibrated and displayed as an indication of the rate of climb or descent in thousands of feet per minute. The main instrument error is the lag in recording because of the nature of the design; for example, when the aircraft levels after a prolonged climb the instrument continues to show a climb until the capsule and instrument case pressures are equal. This has been largely overcome by modern instruments known as inertial lead vertical speed indicators (ILVSI). The VSI is useful in judging climb and descent profiles and is also useful on the final approach stage of the flight, especially when glide slope information is not available. A rate of descent of approximately 800 feet per minute establishes a good final approach descent profile.

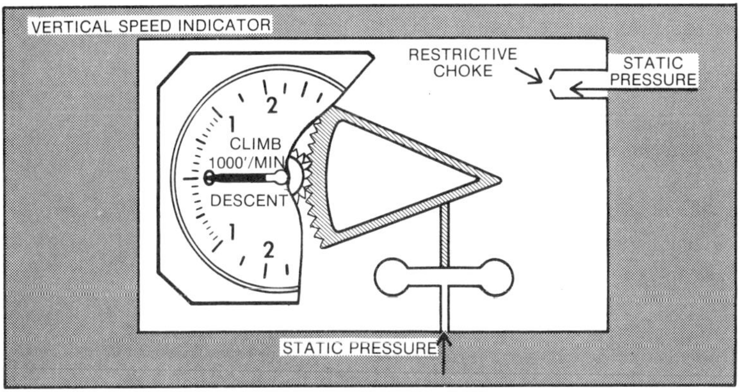

Fig. 6.9 Simple vertical speed indicator.

Instrument flying

One of the most important parts of airline pilot training and practice is instrument flying. Large jet aircraft cannot be flown just by looking out of the window, or by the seat of one's pants (although both help) and all flying, even in clear weather, is by instruments. Of course, on many occasions the pilots do look out for visual reference on a clear day, but all speeds, heights, headings and attitudes are flown by instruments, and the pilot has continually to scan the main flight instruments to prevent attention being focused on any one (Fig. 6.10). No aircraft remains in stable flight for long: speed, height, and heading frequently wander slightly because of gusts, or changes of wind direction, etc., and small control movements are continually required by the pilot (or autopilot) to maintain the desired flight path.

In turbulence the only valid instrument is the artificial horizon, as the others fluctuate wildly, and during such times the pilot concentrates on flying pitch attitude, the only time the aircraft is flown by reference to one flight instrument.

The pilot's most important annual check is the instrument rating which involves a full instrument detail on the simulator (*see* p. 234), including take-off and landing in simulated adverse weather conditions. Faults and failures are fed in along the way, and normally an engine failure is introduced necessitating a three-engine instrument approach and landing. The instrument test must be conducted within the limits laid down for maintaining height, heading and speed, and on the approach the instrument landing system (ILS) limits must not be exceeded. Any deviation outside limits results in an automatic fail, and the pilot is not allowed to fly until the test is retaken and passed.

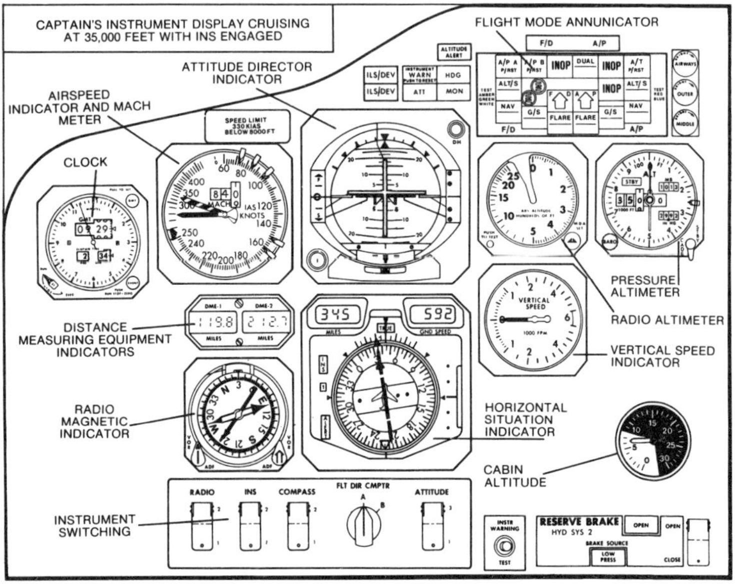

Fig. 6.10 Captain's instrument display when cruising at 35,000 ft with INS engaged.

Spacial disorientation

One problem to be overcome during initial instrument flying training is the effect of a phenomenon known as spacial disorientation, which results from a conflict of the balance senses. Balance is a function of information transmitted to the brain from three separate sources; the eyes, the muscles, and from a section of the middle ear known as the vestibular organ. When flying in clear weather conditions the eyes are the primary source of balance information, and reaction from the other sources is suppressed. Once in a cloud, however, visual reference is lost, except from the instruments, and the pilot becomes more aware of muscular and vestibular senses. Muscle sense arises from skin and joint tension when the body is displaced from the vertical, and vestibular sense from the vestibular apparatus, which consists of three semicircular canals set at right angles to each other, and a static organ. Information supplied to the brain from these sources is a useful supplementary balance aid when on the ground, but when in the air the forces experienced in flight convey to the brain sensations that are often quite different from the information received through the eyes from the instruments. The mental conflict can in extreme cases result in the pilot losing control of the aircraft.

Muscle sense can be confused by accelerations and sharp movements during turbulence, and vestibular sense can be confused by the forces experienced in balanced flight. The three semicircular canals of the vestibular organ positioned at right angles to each other are filled with a liquid with fine hairs projecting into the ends. Change of direction in pitch, roll, and yaw is sensed by movement of the liquid in the canals deflecting the fine sensory hairs. The canals are all connected to a liquid-filled chamber known as the common sac in which is situated the static organ. The static organ consists of small sensory hairs that project vertically upwards and on which lie small crystals of lime salt. Tilting of the body causes the sensory hairs to be deflected and a sensation of tilt is transmitted to the brain. As an aircraft is banked this sensation is experienced by the pilot. However, in a sustained turn, with the in-flight forces in balance, the sensory hairs return in line with the body and the pilot has the distinct impression of being upright and flying straight and level, although his eyes tell him through the instruments that the aircraft is banked in a turn. When the aircraft returns to straight and level flight, the pilot has the distinct impression of banking in the opposite direction until the sensory hairs once again return in line with the body. The sensation can be disturbing, and the conflict of senses quite confusing to the inexperienced instrument pilot. Training is required by the pilot to disregard in flight these supplementary sensations on which he has relied for so long, and to accept and respond only to the numerous flight instrument indications.

Achieving a high degree of accuracy in instrument flying requires a great deal of training, skill, and experience, and frequent practice is necessary to maintain the high standard required for airline flying.

Chapter 7
The Jumbo Flight Deck

The pilots' main instrument panel consists of the Captain's instruments on the left hand side, and a similar layout for the Co-pilot on the right. Below each panel instrument switching allows cross transfer of information. The pilots' central panel contains the main engine instruments, with the standby horizon on the left by the Captain and, below, a warning light system referred to as the annunciator panel. On the right of the engine instruments are the flap position indicators and the gear lever. Along the top of the glare shield is found the autopilot switching, and on either side the VOR/ILS radio selectors. At the top of the central window posts sits the standby compass on its own.

The pilots' overhead panel contains the engine fire handles, engine ignition switches, alternate gear and flap operating switches, engine

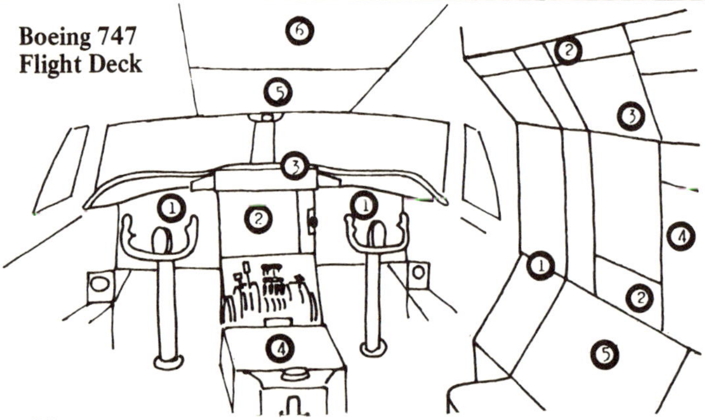

Boeing 747 Flight Deck

Pilots

1. Flight Instruments
2. Main Engine Instruments
3. Autopilot Controls
4. Centre Console
5. Overhead Panel
6 Circuit Breakers

Flight Engineer

1. Electrics & Secondary Engine Instruments
2. Air Conditioning and Pressurisation
3. Pneumatics
4. Hydraulics
5. Fuel

Fig. 7.1 Boeing 747 flight deck. (Courtesy Boeing Airplane Co.).

and airframe anti-icing switches, HF radio selectors, light switches, and other ancillary equipment such as cockpit voice recorder and window heat switches. The pilots' central console contains the three INS sets, VHF radio and ADF selectors, master radio selector boxes, weather radar controls and, at the rear, the rudder trim wheel. On a pedestal on the console are situated the thrust levers, with manual trim and speed brake levers to the left and flap selector lever to the right. Each knob, selector and lever is a different shape, with which the pilots become familiar, and is so designed to help prevent inadvertent operation of the wrong control.

The Flight Engineer's Station.

The Flight Engineer's station contains a mass of instruments covering the many systems on board. To the left of the panel are the auxiliary power unit controls and electrical power switching and indicators. Oxygen, air conditioning, pneumatics, and pressurization controls lie central, with hydraulic indicators, fire protection switching, and numerous warning light systems on the right. Below are secondary engine indicators, the main fuel panel with pump switching and fuel quantity gauges, and the fuel jettison panel. Other ancillary equipment such as engine vibration monitoring, humidifiers, and master radio selector box is situated on the panel. To the right of the Flight Engineer's station, and on the flight deck roof, are found many circuit breakers that protect the numerous electrical systems on the flight deck.

The pilots' instruments and controls.
The pilots' main panel flight instruments have been covered in the previous section, and engine instrumentation in 'The Jet Engine'. Control column operation, central console equipment, and overhead panel switching are discussed mostly in 'Principles of Flight', 'Radio and Radar', and 'Navigation'. The autopilot switching on the main panel glare shield has not been previously discussed and is outlined here.

Autopilot
The autopilot was invented by an American named Sperry, and first demonstrated in Paris as early as 1914. Basically the autopilot consists of a gyro (in the case of the Boeing 747 the INS gyro-stabilized platform) that senses any unscheduled aircraft movement and maintains stable flight. Gyro signals are relayed via a computer to servo motors, which respond to the electrical commands and apply the required hydraulic input to the flying controls. The autopilot also responds to signal demands as selected by the pilot, and as the control surfaces move to commands from the autopilot the control column moves in sympathy. Any trim required is applied automatically.

The autopilot can do no thinking for itself and merely relieves the pilots of the physical handling of the aircraft, which becomes tedious over long periods, and releases the pilots for the more important 'flight management' aspects of the operation. All flight requirements (climb speed, cruise height, heading, etc.) have first to be selected, and the autopilot simply responds to the pilot's demands by flying the aircraft accordingly. The autopilot, therefore, is only as good as the information fed to it: if it is given the wrong information it will fly the aircraft beautifully, without hesitation, in completely the wrong direction. However, the autopilot can fly the aircraft better than the pilots and, of course, does not suffer from fatigue, but even a simple demand to climb the aircraft to a certain height requires all the switching to be made by the pilot (Fig. 7.2).

The autopilot switching is complicated, even for experienced pilots who are new to the Boeing 747, and cannot be covered here. Suffice it to say, therefore, that autopilot operation involves complex procedures that are far from the simple 'push button' task that most imagine. Autopilot engagement varies from aircraft to aircraft, some having the facility for automatic flight from take-off to landing, while others require selection of the autopilot in the climb and disengagement on the approach. The Boeing 747 autopilot is normally engaged at about 10,000 feet, after climb speed is achieved, is used continuously throughout climb, cruise and descent, and is

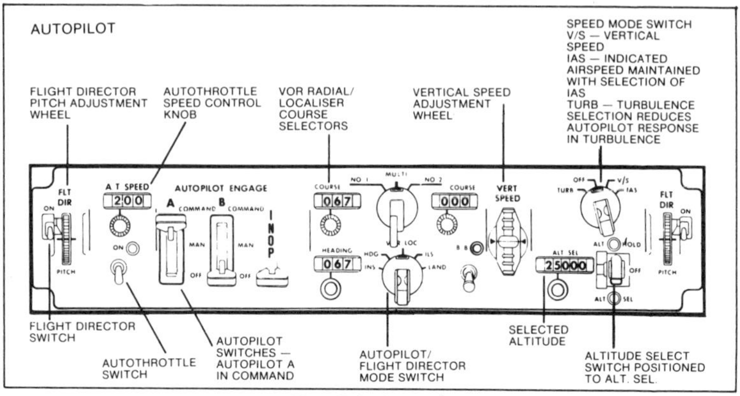

Fig. 7.2 Autopilot.

disconnected at about 3000 feet for a hand-flown approach. Only one autopilot is engaged for normal operations, but two or three are required for automatic landings.

Autoland

Autoland, or automatic landing, not only requires sophisticated autopilot equipment on board the aircraft, but also refined ground facilities at landing airports. The instrument landing system (ILS) ground installation is an integral part of autoland and requires upgrading for automatic landing. Approach and runway lighting systems also require improvement. Where such ground equipment is not available autolands cannot be performed. Straight-in ILS approaches are also required for automatic landings. On the south-east runways at New York, Kennedy, for example, such approaches are not possible because of conflicting traffic from La Guardia to the north, and on the south-east runway at Hong Kong mountainous terrain restricts the approach, thus prohibiting the use of autoland in both cases. The result is that only certain runways at major airports throughout the world are equipped for autoland, and only a handful of these have the facilities for completely blind landings. Autolands are also designed for still, foggy conditions and cannot cope with cross-wind gusts above 15 knots. In such conditions hand-flown approaches are mandatory.

ILS signal quality is defined as category one, two, or three (Cat I, II or III), with Cat III being further divided into a, b, c, Cat IIIc being of the highest quality. An ILS defined as Cat I is not autoland approved, although an autocoupled approach can be attempted whereby the autopilot automatically captures the ILS, but must be disengaged at decision height for a manual landing. Autoland is approved for both Cat II and III approaches, with two and three autopilots being required respectively. The minimum International Civil Aviation Organisation (ICAO) category limits for visibility and decision height are shown in Fig. 7.3, although automatic landings

	CAT I	CAT II	CAT IIIa	CAT IIIb	CAT IIIc
MINIMUM VISIBILITY (METRES)	800 M	400 M	200 M	50 M	0
DECISION HEIGHT (FEET)	200 FT	100 FT	0	0	0

MINIMUM CLOUD BASE HEIGHTS (CAT I – CAT II)

FOG (CAT IIIa – CAT IIIc)

ROLL OUT GUIDANCE REQUIRED (CAT IIIb – CAT IIIc)

COMPLETELY BLIND LANDINGS- ROLL OUT GUIDANCE AND GROUND RADAR REQUIRED (CAT IIIc)

Fig. 7.3 ILS signal quality.

are often carried out in clear conditions for practice. Most Boeing 747 autolands in earnest at the moment are conducted in 'see to land' conditions, with a minimum acceptable visibility of 200 metres and a radio altimeter decision height of 20 feet. At this stage the aircraft is deep in the flare and is in the correct nose up attitude for go-around if insufficient runway is in view. Go-around is automatically initiated by pressing the go-around button and by simultaneously advancing the thrust levers to go-around power.

On aircraft with more sophisticated automatics, Cat III ILS localiser signals provide centre line guidance on roll out to maintain the aircraft straight while decelerating down the runway. When visibility is below 50 metres airport ground radar is required for taxi guidance from the runway to the arrival gate, and to direct fire and emergency services in the event of an accident. At the moment very few airports have such ground radar equipment.

Autoland operation in Cat III conditions requires the crew to establish that the airport is equipped for such approaches, that the aircraft has the capability, and that the pilots are suitably qualified. On initial approach only one autopilot is engaged with the aircraft being steered using the heading control knob. Autothrottle is also engaged with desired speed selected. The decision height of 20 feet is set on both radio altimeters. Radar control positions the aircraft 10-12 miles out to intercept the ILS localiser beam, on a heading angled about 40° to the runway. Both ILS receivers are tuned to the required frequency and the localiser course (i.e. the magnetic direction of the runway) set in both course windows. The autopilot mode selector is then positioned to 'LAND' and the remaining two autopilots engaged. A flight mode annunciator (FMA) indicates autoland selections, and green lights also illuminate with each phase capture. When failures occur, amber or red lights on the FMA illuminate with a steady light, or flash, depending on the severity of the condition.

Flight from now until landing is automatic, but intense monitoring by the crew is required in following the progress of the flight through the various stages to landing. The Captain maintains hands lightly on the controls throughout the approach, ready at a moment's notice to snatch control from the autopilot in the event of failure. At localiser capture the 'NAV' lights on the FMA illuminate green and the aircraft turns onto the runway direction and automatically adjusts for drift. At this stage 20° of flap is set. The gear is selected down with movement of the glide slope (G/S) indicator, and the landing check completed. At glide slope capture the G/S green lights illuminate on the FMA and the aircraft is now fully established on the ILS with all checks complete except landing flap.

Over the outer marker the stop watch is started and time to touch-down noted. Full flap is now selected and speed adjusted to final approach speed. At 1000 feet the triple autopilot operation captures and the flare condition indicates 'armed' on the flight mode annunciator (FMA). With three autopilot channels now in command a system of 'mid-term voting' is employed whereby the aircraft is controlled by the autopilot with the median signal, and, as the approach progresses, control changes from autopilot to autopilot because of small differences in the autopilot command signals. Radio altimeter indications are monitored and at 1000 feet a final 'look around' is made to confirm all checks complete. Clearance to land is received.

The flight deck quietens as tension increases with the approach of landing. Below 500 feet any significant deviation from the ILS is indicated by ILS deviation lights, which illuminate bright red. If the lights come on below 200 feet a go-around must be initiated. The radio altimeter reading unwinds as the aircraft continues descent . . . . 500 ft . . . . 400 ft . . . . 300 ft. The flight directors are switched off at 300 feet. Outside the ground is still obscured by fog. At 200 feet 'alert height' is called by the Co-pilot. The Captain's eyes dart from instruments to outside searching for the first glimmer of a lead-in light, while the Co-pilot's attention remains fixed inside, monitoring the instruments right down to landing. At 100 feet the aircraft is seconds from touch-down. The Captain's forward view is restricted by the instrument glare shield and the reduced slant visual range, allowing only two or three bars of hazy lead-in lights to be visible from the flight deck through the fog. Approaching decision height an aural warning tone sounds and increases in volume with further descent. At 50 feet the glide slope signals are rejected and the autopilot flares the aircraft. The fuzzy green line of threshold lights flash beneath the nose. 'Flare green' is called by the Co-pilot as the respective green light illuminates on the FMA. If 'No flare green' is called, the Captain has to quickly disengage the autopilot and flare the aircraft before it flies onto the runway. At 35 feet the thrust levers close with a distinctive movement. Twenty feet and the aural warning is silenced and an amber decision height light glows — 'Decision height'. The aircraft is deep in the flare. Six or seven runway centre line lights are seen ahead by the Captain — 'Continuing'. (If only two or three lights ahead are seen visibility is much less than 200 metres, requiring a go-around.) Almost immediately the 16 main wheels touch down, followed by the nose wheel lowering gently to the runway. The aircraft is down. The Captain disconnects the autopilot and takes control, selecting reverse thrust and guiding the aircraft along the runway as it decelerates to taxying speed.

The Flight Engineer's instruments and controls.
Aircraft today are extremely sophisticated machines, and only a brief outline of systems can be attempted here. Basically systems are designed on a 'belt and braces' principle with back up facilities available in the event of failure.

Electrics
Four generators (one mounted on each engine) together produce enough electricity (115V a.c.) to supply a small town, such power being required to operate the vast amount of electric, electronic, lighting, and galley equipment aboard the aircraft. Variation in engine speeds with different power settings requires generator constant speed drives and control units to maintain the power supply constant. With an engine or generator failure, switches can be operated to spread the load among the remaining generators. On the ground the power can also be supplied by a ground power unit (GPU), or by the auxiliary power unit generators *(see* Jet Engine p.41). Certain equipment requires d.c. power, and four transformer rectifier units (TRU) convert a.c. power to 28V d.c. Batteries are available to supply essential instruments and controls and to power emergency lighting in the event of complete power failure. The APU also has its own battery for starting, and INS units their own standby battery system. Electrical circuits are protected by circuit breakers which break the circuit by 'popping out' during circuit overload.

The Electrics Bay with racked equipment situated below the first class lounge.

Ground Power Unit.

Air conditioning and pressurization

An aircraft in cruise is like a submarine in reverse! Where the submarine's hull is built to withstand enormous underwater pressures attempting to crush the structure, the aircraft fuselage is designed to contain the pressurized air of the cabin from bursting outwards into the rarefied atmosphere. As the aircraft climbs cabin pressure is reduced to ease outward pressure on the hull and increase aircraft life. Flying at 35,000 feet, for example, the cabin pressure is reduced to an equivalent altitude of 6000 feet, resulting in a differential pressure (the pressure difference between inside and outside the cabin) of 8 pounds per square inch (p.s.i.). (Aircraft passenger doors first move inwards before being swung out to open, so are effectively locked shut by air pressure when the cabin is pressurized.) The 6000 feet environment is not unpleasant for passengers, but those used to living at sea level may feel a little out of breath when moving about the cabin.

The cabin is pressurized via a pneumatic duct which is supplied with highly compressed air bled directly from each jet engine compressor. (On some aircraft there is a separate compressor for cabin air.) The hot, highly compressed air within the pneumatic duct is not only used for air conditioning, but as a source of compressed air for other services. The temperature and pressure within the pneumatic duct are too high for the air to be passed directly to the cabin, so compressed air is first tapped from the duct and fed to three large air-conditioning packs (like giant fridges) which reduce the temperature and pressure to manageable levels. From the packs air flows through a common air-conditioning manifold to four separate cabin zones. To improve circulation, air within each zone can be recirculated by fans. Conditioned air is also channelled behind instrument panels to reduce heat generated by electrical equipment. The air supplied to the cabin is very dry, and humidifiers (which spray a fine mist of water into the system) are sometimes fitted to moisten the atmosphere. However, in spite of such measures, dehydration for passengers on long journeys is a problem. (The dryness of the air can be seen from the way in which bread hardens after a few minutes' exposure to the cabin atmosphere.)

Air from the conditioning packs enters the cabin at a relatively constant mass flow, and pressure is electronically controlled by two outflow valves positioned at the rear of the aircraft. Opening the valves exhausts air to the atmosphere, reducing pressure, and closing increases pressure. In cruise the valves achieve the required pressure while maintaining airflow through the cabin by positioning a little more than three quarters closed. Situating the outflow valves at the rear of the aircraft results in air flowing from nose to tail through the cabin, and smokers are often seated in aft sections to avoid discomfort to other passengers. (Tar from cigarette smoke can be sufficient to jam air filters.) During climb the valves adjust to climb the cabin at an equivalent rate of 500 feet per minute, and in descent at 300 feet per minute. Normally, when descending from cruise level, the cabin altitude is decreased to the height of the destination airport, but on landing in such places as Mexico City, the cabin altitude is actually *increased* (from around 6000 feet to 7300 feet — the height of Mexico City) as the aircraft descends.

An oxgyen supply is maintained on board in the event of a malfunction resulting in depressurization. Masks drop automatically from the cabin ceiling to supply sufficient oxygen to the passengers (the flight crew don special face masks) until an emergency descent can be made to an altitude at which pressurization is not required (normally between 10,000 and 14,000 feet).

Pneumatics

Pneumatic power is used to operate certain equipment, the source of compressed air being the pneumatic duct fed by the engine main compressors. Leading edge flaps arranged in four separate sections on each wing are operated by pneumatic drive units. Sections 2 and 4 on each side extend automatically with selection of 1° of trailing edge flap, and 1 and 3 with selection of 5° of trailing edge flap. In the event of pneumatic failure the leading edge flaps can be extended electrically. Thrust reverser units are also operated pneumatically,

The Nose Wheel Manual Wind-down Facility situated below the first class lounge, forward of the electrics bay.

and air-driven pumps are available as back up for hydraulic system pressurization. The hot compressed air of the pneumatic duct can also be fed to engine nacelles and wing leading edges for anti-icing. On the ground the pneumatic duct can be pressurized from a ground start truck, or from the auxiliary power unit (APU) compressor, and

The Landing Gear Lever on the flight deck.

the system can be used in reverse to feed compressed air to an engine starter motor to turn the engine over for start-up. (*see* Jet Engines p.33).

Hydraulics

Four completely separate hydraulic systems power the flying controls, flaps, speed brakes, stabilizer, landing gear, brakes, and steering. Operation of a particular control on the flight deck, e.g. selection of landing gear lever (up or down), results in forces being transmitted by the virtually incompressible fluid down hydraulic pipe lines to activate equipment. Each system is pressurized by a mechanical engine-driven pump, with a standby air-driven pump (turned by compressed air from the pneumatic duct) available to boost pressure, and to act as a back up in the event of mechanical pump failure. Even with engine failure, a windmilling engine can drive the mechanical pump fast enough to supply pressure to the system, although the air-driven pump operates automatically if pressure drops below a certain level. The flight controls, flaps, landing gear, etc., are each supplied with a mixture of hydraulic systems to reduce the effect of failure of any one system. No. 1 hydraulic system operates flying controls, body and nose gear, inboard trailing edge flap, steering, and secondary brake system. Nos. 2 and 3 hydraulic systems operate flying controls and stabilizer, with No. 2 supplying reserve brakes. No. 4 system operates flying controls, wing gears, outboard trailing edge flaps, and the primary

The gear consists of 16 main wheels set in four bogies of four wheels each.

brake system. Hydraulic systems 2 and 3 are primarily used to power flying controls, but redundancy is such that even if both systems fail all primary flight controls (ailerons, elevators and rudder) can still function with hydraulic systems 1 and 4. Trailing edge flaps can be operated electrically in the event of hydraulic failure.

The landing gear consists of 16 main wheels set in four bogies of four wheels each, with a twin-wheeled axle at the nose. On the ground, while taxying, the aircraft is steered by the nosewheel using a small tiller on the flight deck, and on tighter turns the body gear also steers to aid cornering. Brakes are operated by toe pedals on the rudder bar, but can also be set to apply automatically on landing. Anti-skid units modulate brake pressure when skidding is detected. After take-off, automatic braking occurs during gear retraction to prevent wheels spinning in the bay. In the event of hydraulic failure, the landing gear can be lowered by an alternate system whereby selection of switches on the flight deck operates electric motors to unlock uplatches and wheel bay doors, allowing the landing gear to fall by gravity and air loading to the down locked position. The nose gear has an additional manual wind down facility available if required.

Chapter 8
Meteorology

A full study of meteorology (met.) would fill many books, so this chapter is limited to a simple overall picture of world meteorology, followed by a closer look at some of the localized effects of weather and how they affect pilots.

Weather, with fuel a close second, is the most important factor of any flight; the former often deciding the quantity of the latter, and at pre-flight briefing weather is normally the first item checked. Pilots therefore require a good basic knowledge of met. to study quickly and accurately the charts and information available. The single most important factor, of course, is the forecast weather for the destination airport at the time of arrival of the flight, followed by the forecast for the nominated diversion airport. However, the forecast for the departure airport (in case of a return to base being required), and also forecasts for *en route* airports (which may be required in case of emergency) are also checked. Charts giving such information as upper winds, upper air temperatures, cloud conditions, clear air turbulence (CAT), icing forecasts and so on are also studied and any significant weather affecting the route noted.

Weather at destination and diversion must be above certain minimum conditions, which depend on a number of different factors such as terrain hazards and the quality of radio approach aids. For example, on a flight from London to New York the weather minima, i.e. cloud base height (in feet) and visibility (in metres), at Kennedy International (JFK) must be at or above 300 feet above ground and 400 metres for a hand-flown instrument landing system approach, becoming 200 metres visibility in fog for an automatic landing on suitable runways. Using Boston as diversion, the alternate limits are required to be a little better, and Boston weather must be above 800 feet cloud base and 2 statute miles visibility. (If the system of units seems inconsistent, that's exactly how they are!)

With bad weather forecast before a flight (either at destination or *en route)* alternatives such as a different route, flight level, or even destination (with perhaps extra fuel being carried) are discussed fully before departure, and a decision made by the Captain on any changes required at this stage. However, on all flights, irrespective of weather, frequent listening watches are kept on the radio for the forecasts transmitted from various points throughout the world in case rapid deterioration of weather at destination should occur, or unexpected severe weather be forecast *en route.*

Aircrew, like crews at sea, have a healthy respect for the weather and, again like seamen, treat its fickle ways with caution.

World meteorological conditions

Basically, weather results from fluctuations in temperature and humidity. Since 70 per cent of the world's surface is comprised of water, the source of moisture in the atmosphere caused by evaporation is obvious. Heating, of course, is from the sun. Where the sun's rays strike the earth directly, as at the Equator, the surface temperature is high, and where they strike the earth obliquely, as near the Poles, the temperature is lower, because of the larger surface area to be heated. The Poles, of course, receive the least surface heating, resulting in the formation of large ice caps.

If the earth's surface were level everywhere and the sun's heating uniform throughout the year, it might be supposed that conditions would remain stable. However, the earth's axis of rotation is tilted at an angle to its plane of motion round the sun, with the northern hemisphere being tilted towards the sun in the northern summer, and the southern hemisphere tilted towards the sun in the southern summer. This variable surface heating effect throughout the year results in a constantly changing annual weather pattern. Air is a very poor conductor of heat (i.e. it absorbs heat slowly), so the warmth we feel on earth is a result of the sun's rays heating the surface, which in turn radiates (or throws out) heat to the atmosphere. Land readily absorbs heat from the sun and, in turn, readily radiates heat. In the desert, for example, very high temperatures are experienced during the day, while at night heat loss caused by radiation is extensive, and below zero temperatures can be reached. Similarly, a large land mass like North America can be very hot in summer and very cold in winter. Mountain tops, too, although higher and nearer the sun, are further from the greater levels of heat radiated from the earth's surface and as a result are a lot colder, as evidenced by their snow capped peaks.

Water, on the other hand, is a poorer conductor of heat than land, and is slower to absorb or radiate heat. At the coast this difference in temperature between land and sea during day and night is evident (the land warmer by day, the sea by night) giving rise to the daily frequency of land and sea breezes.

Air masses

The factor most affecting temperature is latitude, and the varying heating effect of the sun at different latitudes results in distinct bands of air being formed round the world, as show in Fig. 8.1.

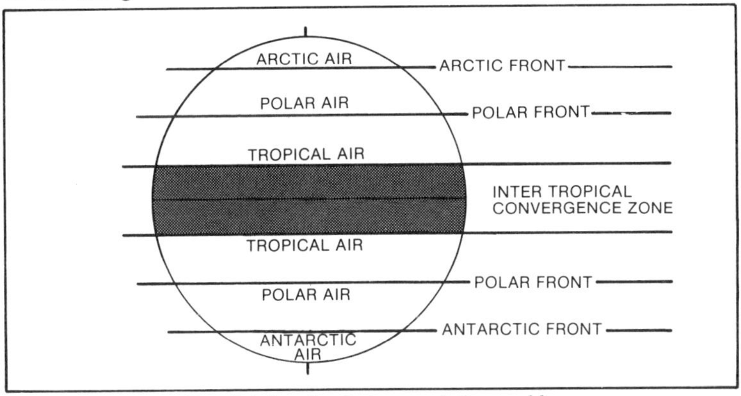

Fig. 8.1 Bands of air round the world.

The property of air within each band depends primarily on the temperature and humidity; the warmer the air, the greater the capacity to hold moisture. When air of more or less the same characteristics covers a large area, perhaps hundreds or even thousands of miles across, it is called an air mass. Although, to assume uniform characteristics, air masses are required to remain relatively stationary for a number of days within a reasonably uniform region, they do not remain stationary for long, and are soon on the move in a constantly changing scene. Air masses are therefore classified according to their source regions. Air masses from source regions within a particular latitude band have a similar temperature range, but the humidity depends on whether the source region is over land or water, and air masses are further subdivided into 'continental' and 'maritime'. For example, one source region of tropical continental air in winter is over North Africa and another in summer over the mid south of the United States. As can be imagined, tropical continental air is warm and dry. Tropical maritime air, on the other hand, has source regions in the South Atlantic and North Pacific, resulting in air of high temperature and high humidity.

Frontal Activity

Divisions between bands of air with different characteristics, for example between the principal bands of Arctic (or Antarctic) air and Polar air, and between Polar air and Tropical air, are well defined, although the transition zone may be several miles wide. Local air masses from adjacent principal air bands move with the wind and infringe upon each other's territories, forming wave-like zones of conflict between the principal air bands. Two Norwegian meteorologists studying this phenomenon during the First World War aptly named the lines of limit of advance of air masses as fronts. The differences between air masses are mostly in temperature and humidity, the most significant, as a rule, being temperature. Where a warm air mass is advancing and displacing a colder air mass, the front line between the two is known as a 'warm front' and where a cold air mass is displacing a warmer air mass the front line is known as a 'cold front'. The transition zone between Arctic and Polar air bands is named the Arctic Front (the Antarctic Front is similar) and between the Polar and Tropical air bands the Polar Front. The transition zone between the Tropical air bands coverging at the Equator is less well defined and is discussed later.

The fronts do not remain stationary but shift with the heating effect of the sun, moving north in the summer and south in the winter in the northern hemisphere, and *vice versa* in the southern hemisphere. An example of intense frontal activity is where the Polar Front lies across the North Atlantic. The interaction between opposing air masses results in a fragmented wave-like pattern of frontal sections being formed, stretching across the Atlantic in family groups from Florida to the south-west of Britain in the winter, and from New Foundland to the Faroe Islands in the summer, drifting continuously eastwards towards Europe and Scandinavia on the prevailing westerly winds. The fronts can travel at speeds between 30 and 40 knots, and can easily cover 1000 n.m. in one day. Fronts seen in the middle of the Atlantic often reach Europe within twenty four hours.

Although temperature differences between the principal air bands of Arctic (or Antarctic) and Polar air, and between Polar and Tropical air are sufficient to produce a distinct frontal structure, the same is not true where the Tropical air bands north and south converge at the Equator. In this case each possesses almost identical characteristics of temperature and humidity, and the transition zone between the two is broad and ill defined. This convergence zone of the Tropical air bands can be hundreds of miles wide, is too diffuse to be recognised as a distinct front, and is termed the intertropical convergence zone, or ITCZ for short. The ITCZ also shifts with the influence of the sun, lying due north of the Equator in the northern summer and mostly south of the Equator in the southern summer. Air masses within the convergence zone are generally warm, moist and unstable, resulting in intense weather activity over a wide area. The ITCZ traverses large areas of ocean on its northwards passage and the warm air becomes heavy with moisture. Vast cloud formations form with towering thunderclouds 200-300 miles deep, producing frequent storms and heavy tropical downpours.

The severe S.W. monsoon weather in India and the Far East is a result of the movement of the ITCZ and first appears in early summer as the ITCZ passes overhead on its northwards journey. Southwards movement of the ITCZ occurs with the approach of winter, which results in the N.E. monsoon, but the air is now drier and more stable with the ITCZ approaching mostly from over land. ('monsoon' refers to the wind that accompanies the ITCZ and is derived from the Arabic for 'season'.) Movement of the ITCZ and its associated severe weather conditions are well known to all pilots operating within its sphere of influence, and it is no coincidence that most flight crew prefer to go anywhere rather than India in early summer.

Although frontal activity is mostly evident at the transition zones of the principal air bands, the picture is more complex as frontal activity also occurs between air masses from different sources within a particular air band, or even between air masses from the same source meeting with some difference in characteristics after taking different paths.

Distribution of pressure and wind

The heating of the sun, and therefore the temperature of the air, also has an influence on pressure. Pressure is the effect of the weight of air acting on the surface of the earth; where the air is warmed it is lighter and less dense because of expansion, resulting in lower surface pressure. Where the air is cold it is heavier and more dense, and surface pressure tends to be higher.

In general terms, therefore, the distribution of surface pressure is such that in warmer, less dense air, as at the Equator, the surface pressure is low, and at the Poles, where the air is cold and more dense, the surface pressure is high. The geographic distribution of oceans and land on the surface of the earth and the variable heating effect this causes results in alternate bands of high and low surface pressure extending round the earth, as shown in Fig. 8.2. As can be seen, in winter the low pressure areas tend to develop over the warmer oceans and high pressure areas over the colder land masses, while in summer highs tend to develop over the cooler oceans and lows over the warmer lands.

A surface pressure chart, if shown in three dimensions, would be seen to rise and fall like the contours of the earth, and, indeed, pressure patterns are described similarly to undulations on an ordance survey map, e.g. troughs and depressions for low pressure areas and cols and ridges for high.

Air flows from regions of high pressure to regions of low pressure.

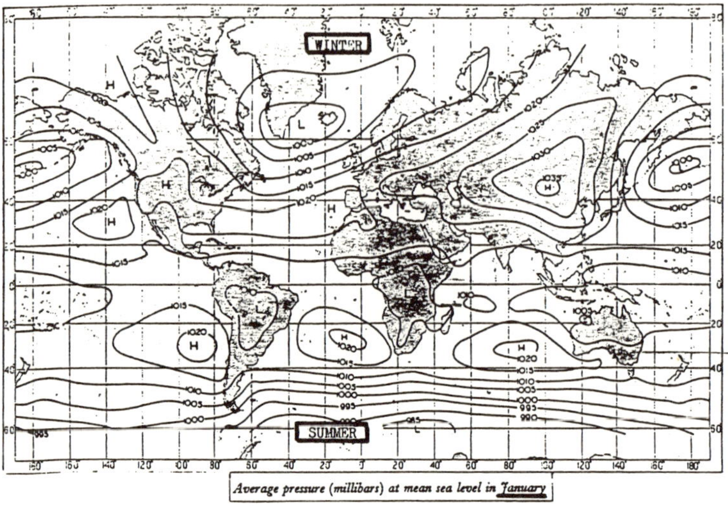

Fig. 8.2 Average surface pressures, world wide, in January and July.

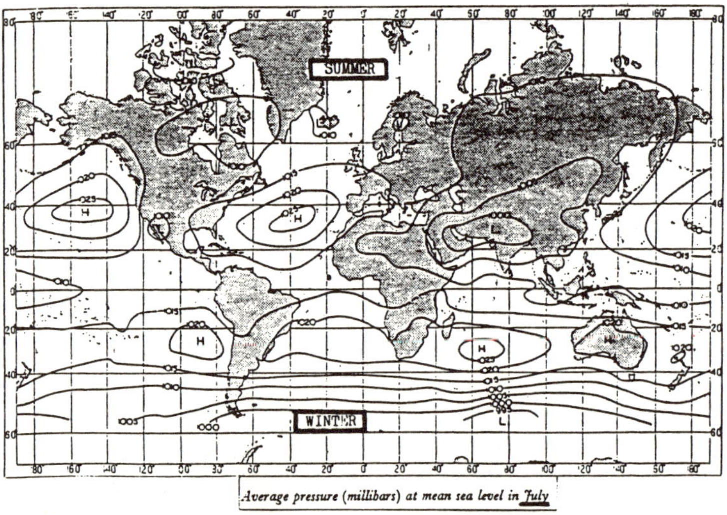

Since the areas are generally well defined, it is not difficult to construct an idealised diagram (Fig. 8.3) showing the distribution of pressure and wind and the general circulation of upper air currents throughout the world, although in reality the situation is more complex.

Air does not flow directly from high to low pressure areas because of the spinning of the earth (the Coriolis effect), which results in

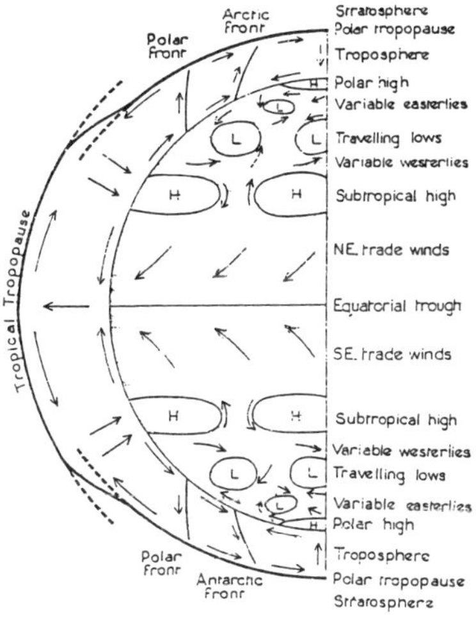

(The Handbook of Aviation Meteorology)

Fig. 8.3 Idealized distribution of pressure and wind.

winds being deflected to the right in the northern hemisphere and to the left in the southern hemisphere. In the northern hemisphere (Fig. 8.4) low pressure areas have a cyclonic motion consisting of a circular wind movement rotating in an anticlockwise direction, and in high pressure areas an anticyclonic motion rotating in a clockwise direction, and *vice versa* in the southern hemisphere. High pressure areas, therefore, are frequently referred to as anticyclones, but low pressure areas are more commonly referred to as depressions. In some tropical regions revolving storms resulting from intense depressions are known as cyclones, but also as hurricanes and typhoons, depending on the area. Localized depressions with strong cyclonic winds are frequent in the United States and are known as tornadoes.

Weather is closely dependent on the distribution of pressure at the surface, and low pressure areas consisting of relatively warmer, moister and more unstable air are associated with wet, windy, unsettled weather, while high pressure areas consisting of cooler, drier and more stable air are associated with clear, calm weather.

The tropopause

One further effect of heating by the sun is that the stronger the sun's rays the greater the depth of atmosphere heated. The atmosphere

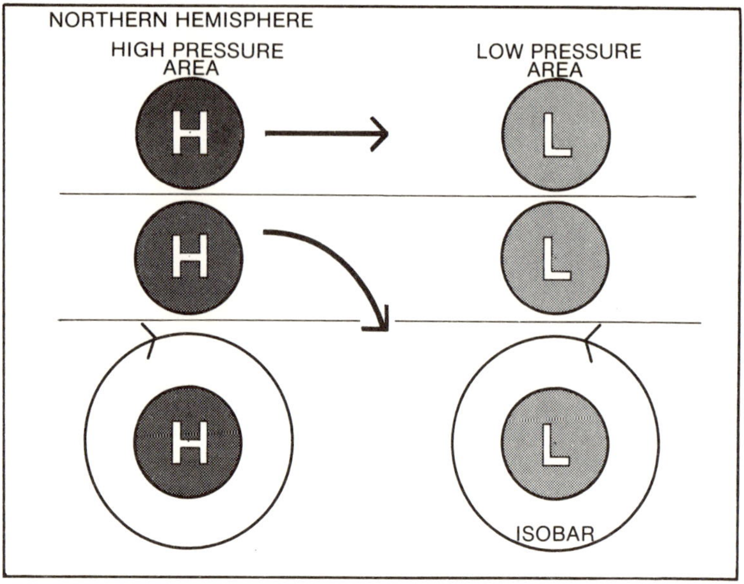

Fig. 8.4 Wind movements in the northern hemisphere.

cools with height at a rate of approximately 2°C per thousand feet, and this cooling continues until a point is reached at which the temperature is considered to remain constant at around minus 57°C. The lower levels of the atmosphere at which this assumed standard rate occurs is known as the troposphere. Where the temperature remains constant (and eventually at greater height begins to rise) is known as the stratosphere. The dividing line between the two is the tropopause. Since warm air rises into the cooler air above, air continues to rise (cooling as it does so) until at the tropopause it ceases to rise further as the temperature above remains constant.

Clouds are the result of moisture within the atmosphere condensing when moist air is cooled on rising, so the tropopause is important to pilots in that it generally marks the height limit of cloud formation. The tropopause also indicates maximum heights of strong winds (*see* Jet Streams p. 159). The depth of atmospheric heating at the Equator is greater than at the Poles and, not surprisingly, the height of the tropopause is greater at the Equator. The height of the tropopause also varies with the seasons, being higher in summer than in winter, except at the Equator where it remains at about 55,000 feet. The average tropopause varies from winter to summer between 30,000 - 35,000 feet at mid-latitudes, and 20,000 - 25,000 feet at the Poles.

The international standard atmosphere

Changes in atmospheric conditions are a result of fluctuations in temperature and pressure (and density which is closely related). Such variation in values required a standard atmospheric condition to be defined as a basis for scientific reference and, in aviation, for the calibration of flight instruments and the measurement of aircraft performance.

Average mid-latitude values were used as a datum and the international standard atmosphere (ISA) was defined by the International Civil Aviation Organisation (ICAO) as a temperature of 15°C (59°F) and a pressure of 1013.2 millibars (29.92 inches of mercury).

In aviation, temperature at height is expressed as a temperature deviation from the ISA temperature. For example, ISA temperature at sea level is + 15°C, decreasing by 2°C per thousand feet, so at 35,000 feet the ISA temperature is -55°C. If the actual temperature indicated is colder than -55°C, say -60°C, then actual temperature is expressed as ISA -5°C at 35,000 feet, and all performance graphs are entered using temperature expressed in this manner.

Pressure is expressed as a force per unit area, the standard meteorological pressure being 1013.2 mb, which is perhaps more familiar when given as 14.7 pounds per square inch or 1.03 kilogrammes per square centimetre. Measurement of pressure can also be expressed as the height of a column of mercury which can be supported by the atmosphere (i.e. a barometer), the standard pressure value being 29.92 inches, or 760 millimetres of mercury. Surface pressure charts display the pressure distribution at mean sea level (MSL — the average level between high and low tide), with lines joining points of equal pressure known as isobars.

Wind and isobars are related in that, as mentioned earlier, winds are deflected in their passage from high to low pressure by the spinning earth, and in general isobars indicate the line of direction of the wind. The strength of the wind is indicated by the separation between isobars, the closer together the isobars, the stronger the wind.

Local effect of weather

The preceding paragraphs have covered the generalized climatology of the world, showing the effects of solar heating and the idealized distribution of surface pressure and wind. However, in the ever-changing weather scene the situation is infinitely more complex, and locality and geography play major roles in climatic effects. Latitude of a position, for example, without reference to its geographical situation, can often give a confusing image of climate. Nairobi, although situated close to the Equator, stands at 5000 feet, and has winters that are cool (and even chilly in the evening) and summers hot but pleasant. New York, although the same latitude as Madrid (approximately 40°N) is affected by the large North American land mass, and as a result suffers bitterly cold winters, while Madrid rarely sees snow. Japan lies even further south (Tokyo is approximately 35°N) and has hot summers, but in winter prevailing winds from the north sweep across the large cold expanse of the Asian land mass and bring cold air and freezing winters. Weather, therefore, varies considerably throughout the world, and the following paragraphs look more closely at certain aspects of weather of interest to the pilot.

Cloud and rain

In the eighteenth century a London chemist and botanist by the name of Luke Howard first classified clouds into ten practical types, using grand Latin descriptions. Later, meteorologists expanded on his work and many different cloud types are now classified, but need not be covered here. Basically, heaped cotton-wool type clouds are called cumulus and layer type clouds stratus.

Clouds at medium height (from 7000 to 20,000 feet) are given the prefix alto, i.e. altocumulus and altostratus, and high cloud (above 20,000 feet) the prefix cirro, i.e. cirrocumulus and cirrostratus. Thin wisps of cloud aloft are known as cirrus, and a mixture of cloud types at low level is known as stratocumulus. Rain clouds are nimbostratus (Ns), associated with prolonged light rain, and cumulonimbus (Cb) with heavier showery rain.

Clouds are formed not only by thermal convection, when moisture condenses with the cooling of rising warm air, but also by lifting, when air is forced upwards over hills and mountains, and also at fronts where cold air masses undermine warm air masses, forcing the warm air to rise. Indeed, extensive cloud formation is associated with frontal activity, and a frontal surface is clearly marked by a distinct line of cloud. Precipitation occurs when the extent of condensation saturates the air and moisture falls from the cloud as rain, snow or hail, depending on the temperature conditions.

In airport weather reports, cloud cover is given in terms of octas (eighths), 8/8ths being complete cloud cover, and cloud base heights are given in hundreds or thousands of feet. In low cloud conditions, cloud base height is important as pilots require to know the height at which the aircraft will break cloud.

Thunderclouds are large towering cumulonimbus (usually simply referred to as Cb), often anvil in shape, and are a distinct hazard to aviation. Within the cloud large water droplets churn up the air as they fall from higher levels and rise again on updraughts at up to 5000

Giant Cumulonimbus (CB) build-up over mid Atlantic.

Weather Radar.

feet per minute, causing severe turbulence. In the vicinity of airports thunderclouds can cause dangerous downdraughts on the cloud fringe and take-offs are normally delayed until they pass. However, accidents associated with thundercloud activity are rare, as aircraft today are very strongly built and pilots make every attempt to avoid such clouds to give their passengers a smooth ride. Weather radar on board can pick up Cb at up to 300 n.m. as the radar signals reflect from the large water droplets within the cloud. Modern radar sets are now in colour with thunderclouds appearing in red and others in green. Clouds with low moisture content do not appear on radar.

Weather radar is invaluable in avoiding Cb activity when aircraft are already in cloud, or at night in areas such as the tropics, where giant thunderclouds can stretch up to 50,000 or 60,000 feet. Friction within the cloud causes an electrical build up, which discharges as lightning with a loud bang (caused by swift and violent heating of the air), and which can damage aircraft if struck. Electrical static on aircraft is a hazard, and tyres are specially treated to discharge electricity to the ground on landing. Static wicks suspended from the trailing edge of wingtips also discharge static electricity to the atmosphere in flight. (See photo p.59). In the vicinity of thunderclouds pilots take avoiding action where possible, by requesting re-routeings or a change of flight level, and pass at least 20 n.m. upwind of Cb to avoid the turbulent wake downwind. On occasions deviation from flight path is not approved because of traffic and the aircraft simply has to weather the storm. Very rarely, with a severe line of thunderclouds that are unavoidable, aircraft may turn back. In turbulent conditions passenger comfort is improved by selecting 'turbulence mode' on the autopilot, which reduces autopilot reaction, and pilots slow the aircraft to a rough air speed of around 300 knots indicated to reduce structural stresses.

Wind

As previously described, wind is the result of the movement of air from high to low pressure which is deflected by the spinning earth.

Buys Ballot, an early meteorologist, formulated a law which states that 'with an observer's back to the wind the low pressure area is to the left in the northern hemisphere and to the right in the southern hemisphere'. The distribution of high and low pressure areas throughout the world creates seasonal winds whose names are well known from the days of sail. The Trade Winds, lying north and south of the Equator, were just that, carrying trading ships to the far reaches of the world. The 'Roaring Forties', winds notorious for their wildness and strength, lie at 40°S, and slack areas of the Equator belt, where winds are calm, are known as the Doldrums. Local winds, too, are named, for example, the 'Mistral' in France, 'Fohn' in the Alps, and 'Chinook' in the Rockies. They are the result of the geographical influence of valleys and mountains channelling the wind, or the variable heating effect of the sun on their slopes. Airports are constructed with local wind conditions closely in mind as runways have to be built into wind.

In the wake of a strong wind blowing over a mountain, large wave currents of air known as standing waves can develop which can be a hazard to aircraft.

The effects of temperature and pressure at altitude are such that air at height flows from a high to a low temperature region. Air over the Equator is warmer than over the Poles so air aloft tends to move north and south from the Equator. Since the spinning earth deflects moving air to the right in the northern hemisphere and to the left in the southern hemisphere, wind at altitude is generally westerly throughout the world (Fig. 8.5). Also, at altitude the effects of temperature and pressure are compounded to produce winds of enormous speed. Where the wind is concentrated into a fast flowing

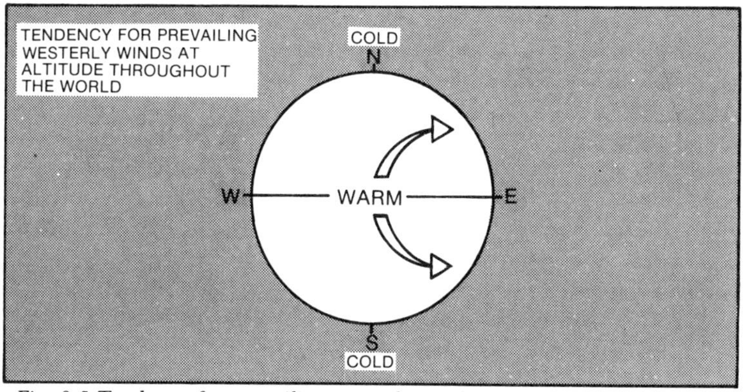

Fig. 8.5 Tendency for prevailing westerly winds at altitude throughout the world.

river of air only a few miles in depth, but perhaps a few hundred miles wide and a thousand miles long, it is known as a jet stream. Wind speeds at the centre can often reach 200 knots, and occasionally even greater. Jets streams are frequently found over the North Atlantic (Fig. 8.6) and lie just below the tropopause at standard cruise altitudes. Flights eastbound are therefore planned to take advantage of the jet streams, and flights westbound to avoid them. Average upper wind components on the North Atlantic are westerly at 60 knots and can add up to one hour on westerly crossings. From Europe to Australia average winds are westerly at 25 knots to the

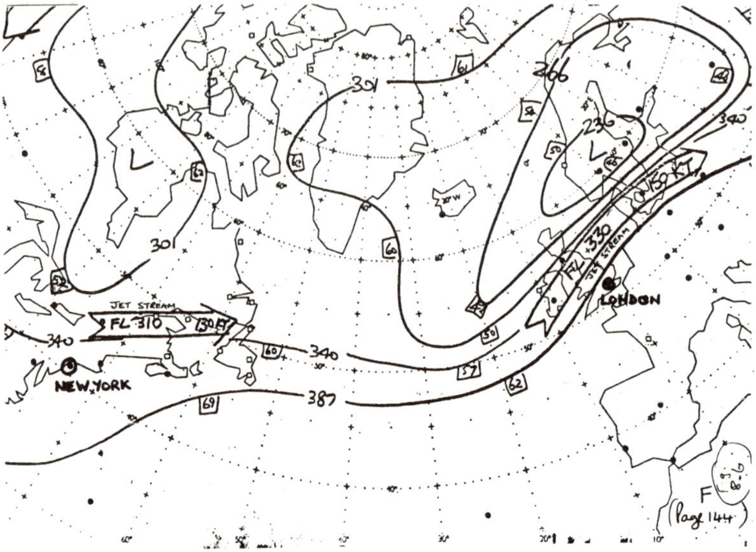

Fig. 8.6 Jet streams over the North Atlantic.

(Courtesy of Met. Office, Bracknell, England).

Middle East, and from the Middle East to India westerly at 60 knots. From India the westerly winds decrease towards Singapore, and from there to Darwin become easterly although light. From Darwin to Sydney the average wind component is westerly at 60 knots.

Clear air turbulence (CAT) is a result of windshear, where the wind changes in strength from one point to another. A wind change of as little as 4 knots per thousand feet can cause bumpy conditions and 6 knots per thousand feet severe turbulence. Certain charts mark these areas by numbers related to the wind shear in knots per thousand feet and are an indication to pilots of the probability of turbulence. However, forecasting is difficult, and frequently smooth rides are experienced where turbulence is expected and *vice versa.* At the moment, detecting CAT in flight is not possible, although rapid outside air temperature (OAT) changes can be an indication of imminent turbulence. It is advisable, therefore, to maintain seat belts fastened while seated, as unexpected CAT can result in passengers being injured in the cabin.

Fronts

Worldwide distribution of frontal activity was covered earlier, and here the formation of fronts and associated weather conditions are examined more closely. As an example, activity along the Northern Polar Front shows clearly the interaction between opposing air masses (Fig. 8.7).

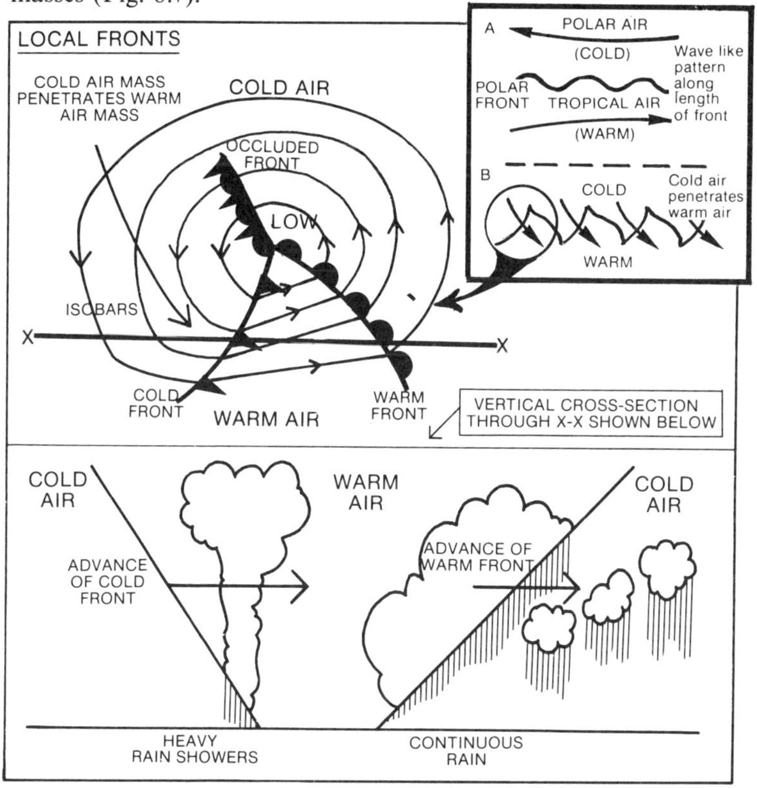

Fig. 8.7 Local fronts.

Movement of adjacent air masses sets up a wave-like pattern along the length of the front, and the cold air penetrates the warm air mass and advances into the warm air. The cold salients eventually merge (or occlude) forcing the warm air upwards and forming an 'occluded front' while the pressure falls and a depression develops. The fronts eventually fragment, scatter and drift in the general direction of the wind, and at the same speed. A frontal depression and the distribution of isobars is shown in Fig. 8.7 and a chart of significant weather (Fig. 8.8) clearly shows the activity of the Polar Front on the North Atlantic.

The front lines of the warm and cold air masses indicate the line of advance of the front on the surface. However, the cold air wedges below the warm air forming a sloping surface and the front line

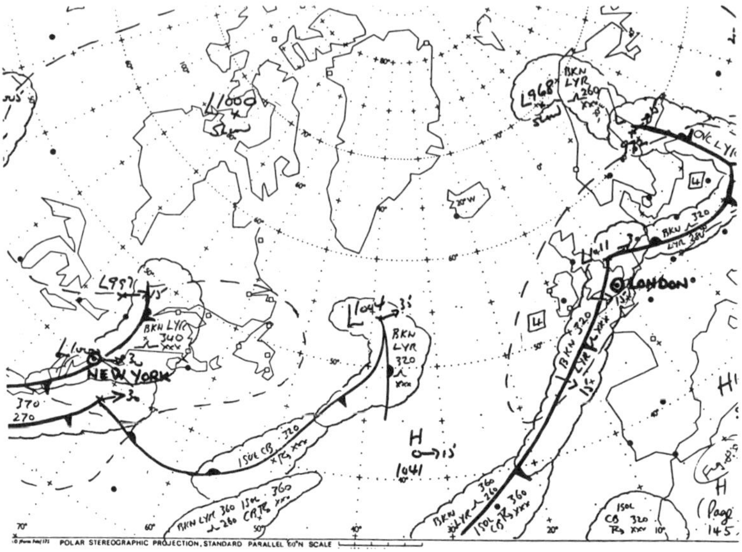

Fig. 8.8 Chart of significant weather over the North Atlantic.
(Courtesy of Met. Office, Bracknell, England).

marks the line at which the frontal surface meets the earth. A cross section through the frontal surfaces at X - X is shown in end view below. At both fronts it is the warm air that is being lifted, forming cloud and producing precipitation, but the slope of the frontal surface is steeper at the advance of the cold front than at the warm. With the approach of a warm front the wind and temperature increase slightly, the barometer falls, and drizzle or continuous rain commences. With the passage of the front, the wind direction changes, temperature rises and the rain slackens or stops. At the approach of a cold front the wind increases, the barometer falls, but the temperature remains steady with perhaps some showers. With the passage of the front the wind increases and changes sharply, the barometer rises suddenly and the temperature falls. Heavy rain showers occur, with possibly thunder and hail.

Weather forecasting is closely dependent on the analysis of the movement of fronts, and on calculation of the direction in which they travel. Unfortunately, fronts are fickle and do not always move as expected; they may suddenly change direction or be temporarily halted by the advance of a high pressure region. Not surprisingly, in spite of sophisticated techniques accurate weather forecasting can still be a difficult task.

Visibility

Visibility at airports is important to pilots in that it is still essential, on almost all occasions, to be able to see a certain distance on landing, and visibility reports at airports are given in metres, kilometres, or (in the USA) in statute miles. When visibility is poor, actual visual range along a runway is measured by instruments known as transmissometers, which detect transparency of the atmosphere, and is quoted as runway visual range (RVR) for that particular runway; e.g. RVR Runway 26, 400 metres.

Visibility can be reduced by dust or smoke particles producing haze. Rain or low cloud create misty conditions, and where moisture in the atmosphere reduces visibility to less than one kilometre, fog is considered to exist. A mixture of fog and pollutants produces the kind of 'pea soup' fog (or smog) familiar to Londoners in the fifties before the Clean Air Act.

One common phenomenon is low-lying mist or haze, which reduces visibility on landing. At altitude it is possible to see the ground clearly when looking vertically down through the shallow layer, but on the approach, when viewing diagonally through the layer, visibility can be markedly reduced and may even be below limits for landing.

In India, near the cities, the occurrence of shallow smoke haze in the morning has just this effect. The haze actually clears quickly with the sun, and when visibility is reported below limits, aircraft normally circle overhead the airport at height awaiting an improvement. Holding overhead Bombay, for example, it is difficult to explain to passengers that the aircraft is unable to land because of bad visibility when they can see the ground so clearly. Passengers are usually quite surprised at the reduction in visibility when the aircraft descends into the shallow haze layer, but the phenomenon is well known to pilots.

Fog at airports is not the problem it used to be, although thick fog still closes airports in spite of the sophisticated equipment available. Regular use by all aircraft of blind landing and taxying techniques in dense fog conditions is still some way off.

On take-off, aircraft like the Boeing 707 and Boeing 747, with engines slung in pods below the wings, are subject to a marked swing if an outboard engine fails and therefore require better visibility for take-off in fog conditions than, say, a Boeing 727 or DC10, which have the engines bunched more closely. When dense fog does prevail, more often than not passengers have no choice but to wait it out on the ground or, as they have done since the beginning of flight, seek alternative transport.

Fog is usually the result of water vapour in the atmosphere cooling and condensing near the surface and may be described as a cloud on the ground. As air is cooled, the temperature drops to a point at which the air can no longer hold the water vapour present, and condensation produces moisture. This temperature is known as the dew point. Cooling can occur by radiation to the atmosphere. If the cooling effect is spread by a light wind, the land can cool through the night to a temperature below the dew point of the air, producing radiation fog in the morning, a common occurrence in Europe in winter. (Cloud cover at night can actually prevent fog by insulating against the effects of radiation.) Fog can also form when warm moist air is cooled by moving over a cold surface with a temperature below the dew point of the air. Such fog is known as advection fog and is common in winter in areas like California when warm moist air from the sea spreads over the cooler land.

Dew point temperature is important to pilots because it marks the temperature to which the air must drop before fog can develop. In actual weather reports the air temperature and dew point temperature are quoted. When the air temperature drops to the dew point temperature and the two are quoted as equal, e.g. temperature 8°C/dew point temperature 8°C, the possibility of fog is evident.

The Unwinged Ones

by Ogden Nash
From: *The Private Dining-Room* (Dent, 1953).

I don't travel on planes
I travel on trains.
Once a while, on trains,
I see people who travel on planes.
Every once in a while I'm surrounded
By people whose planes have been grounded.
I'm enthralled by their air-minded snobbery,
Their exclusive hobnobbery,
And I'll swear to, before any notary,
The clichés of their coterie.
They feel that they have to explain
How they happen to be on a train,
For even in drawing room A
They seem to feel declassé.
So they sit with portentous faces
Clutching their attaché cases.
As the Scotches they rapidly drain

That they couldn't have got on the plane,
They grumble and fume about how
They'd have been in Miami by now.
They frowningly glance at their watches,
And order more Scotches.
By the time that they're passing through Rahway
They should be in Havana or Norway,
And they strongly imply that perhaps,
Since they're late, the world will collapse.
Then, as station merges with station,
They complain of the noise and vibration,
These outcasts of aviation,
They complain of the noise and vibration.
Sometimes on the train I'm surrounded
By people whose planes have been grounded.
That's the only trouble with trains;
When it fogs, when it smogs, when it rains,
You get people from planes.

Ice

Ice and snow on taxiways and runways can be a major problem, and take-off weights have to be reduced in such conditions. Every attempt is made at snowbound airports to keep runways clear, but where snow or ice build up on the runway to above a certain depth (i.e. 38 mm dry snow or 13 mm wet snow or slush), take-offs cannot be attempted, although landings are permitted on up to 10 cm of dry snow. Heavier falls result in an airport being closed. Snow lying on wings produces unacceptable aerodynamic qualities and has to be removed before departure by brushing or hosing with a de-icing fluid.

Motorists understand the problems of driving on snowy or icy roads, and handling large aircraft in such conditions is no easy task. All turns on slippery taxiways are reduced to a maximum of 5 knots for the big jets to prevent skidding, and care has to be taken to maintain the runway centre line during take-off and landing, especially in windy conditions.

Flight deck windows and external instrument sensors are heated to prevent icing, and hot air tapped from the engines is available for anti-icing of engine nacelles and aircraft airframe. Engine anti-icing is frequently used as a precautionary measure, but the big jets seldom experience conditions where airframe anti-icing is required. However, serious icing in flight can be a hazard; one example being rime ice, which forms as a result of super cooled water droplets (drops of water in the atmosphere that maintain their liquid state at temperatures below the freezing point) freezing on contact with the aircraft. Where there is little spreading of the water drops on contact, air is trapped between the particles producing an opaque appearance, and where spreading occurs clear translucent ice results. Airframe anti-icing is switched on when such icing is evident and hot air is supplied to the leading edges of wings to de-ice any ice accretion.

Chapter 9
Air Traffic Control

Air traffic control (ATC) is a study and a career on its own and can be discussed here only from the pilot's point of view. Whether pilot or controller, however, the 'Rules of the Air' are the very basics of safe flight and have to be understood by all. To list all the rules would, of course, be laborious, and specific regulations are only mentioned as the need arises. Where a particular 'Rule of the Air' has been stated in another chapter it is not repeated here.

Units of measurement in aviation

The unit system in aviation varies throughout the world and at present the situation can only be described as a mess. The International Civil Aviation Organisation (ICAO) standards are used in most parts of the world, but even these are a mixture: height in feet, speed in knots, wind speed in knots, distance in nautical miles, runway lengths in metres (although often quoted in feet), weight in kilograms, temperature in degrees Centigrade, pressure in millibars, visibility in metres and kilometres, volume in litres (although, in practice, refuelling is occasionally in gallons and, believe it or not, oil replenishment in anything from pints to US quarts, depending on aircraft type). The United States retains a modified 'Imperial' system: height in feet, speed in knots, wind speed in knots, distance in nautical miles, runway lengths in feet, weight in pounds, temperature in degrees Fahrenheit, pressure in inches of mercury, visibility in feet and statute miles, and volume in U.S. gallons. When operating within the USA, aircraft complying with ICAO standards have to convert just about every unit for calculations before departure. The USSR, the Republic of China, and the Eastern bloc countries use the metric system throughout: height in metres, speed in kilometres per hour, wind speed in metres per second, distance in kilometres, runway length in metres, weight in kilograms, temperature in degrees Centigrade, pressure in millimetres of mercury, and volume in litres.

Controlled airspace

All big jet aircraft operate within controlled airspace and so are subject to instrument flight rules (IFR) which stipulate than an air traffic control (ATC) flight plan must be submitted with details of the flight, ATC clearances and instructions must be adhered to, certain appropriate radio equipment must be carried, and pilots must be suitably licensed, i.e. hold a current instrument rating.

Controlled airspace in the immediate vicinity of an airport is known as a control zone (CTR or CTZ — generally from ground level to 3000 feet), and where a zone is extended at a major airport (perhaps to include a group of airports such as Heathrow and Gatwick at London, or Newark, La Guardia and JFK at New York) it is known as a terminal control area (TMA or TCA). The vertical limits of TMA airspace are indicated on charts, e.g. from 2000 feet to flight level 250.

Certain larger areas of high traffic density are designated control areas (CTA), and boundaries are clearly defined on charts with vertical limits of airspace indicated (e.g. Piarco CTA flight level 60 - 200). Control areas above the Atlantic and Pacific oceans are known as ocean control areas (OCA). Airspace above about 20,000 feet (the height varies throughout the world) is known as upper airspace, and control areas at these levels are known as upper control areas (UTA); e.g. France UTA flight level 195 - 660. In many countries all upper airspace is controlled to very high limits (as in the example of France given above) or to an unlimited height.

Criss-crossing the world are the aerial highways known as airways, up to ten nautical miles wide and marked at each end by radio beacons. All airways are designated controlled airspace and vertical limits are indicated on charts, e.g. 3000 feet to flight level 460. The International Civil Aviation Organisation annotates airways by colour and number; e.g. Amber 10 (A10), Green 1 (G1), Blue 15 (B15), Red 3 (R3), White 4 (W4), although this is changing to the simpler Alpha 10, Golf 1, etc. Airways within upper airspace are given the prefix 'upper', e.g. Upper Blue 4 (UB4). In the United States, airways below 18,000 feet are Victor airways and above that height Jetways, e.g. Victor 494, Jet 121.

All airspace throughout the world is sectioned into large regions known as flight information regions (FIR); upper flight information regions — UIR — in upper airspace, and boundaries are clearly marked on charts. The boundary between two FIRs lying in different countries lies along the border line. Within an FIR/UIR, only aircraft operating in controlled airspace as outlined earlier are subject to air traffic control (ATC) and instrument flight rules (IFR). Outside controlled airspace aircraft are usually free to come and go as they please, although they are still subject to to the basic 'Rules of the Air'. However, in many countries, when flying outside controlled airspace even light aircraft journeys over a certain distance (especially in places like Canada with large remote areas), or over water flights of more than a certain distance, require a flight plan to be submitted and are subject to ATC. The days of the 'Freedom of the Air' are not quite what they used to be!

On the North Atlantic, because of the volume of traffic travelling in one directon at a time, a North Atlantic track system operates with a series of approximately parallel tracks, which take advantage of the best winds or avoid the worst. Computer calculations indicate the best routeings, and available tracks are published twice daily. All airspace within the North Atlantic track system above flight level 280 is under ATC from both sides of the Atlantic.

Separation

At present there are no in-flight collision detection devices on board aircraft and pilots rely heavily on ATC to maintain separation between flights. At certain air traffic control centres (ATCC), however 'conflict alert' equipment is available that can indicate potential incidents to controllers. Unfortunately, occasions do arise when aircraft pass unacceptably close and in such cases 'airmiss' reports are filed by the pilots concerned. The main function of ATC therefore, is to maintain safe separation between all traffic. The minimum acceptable limits between aircraft in flight are defined as follows.

At airports under radar control aircraft departing on the same initial track have a minimum separation of two minutes, and on different tracks, one minute. (A light aircraft taking off behind a big jet may have the separation increased to ten minutes to avoid wake turbulence left by the departing jet.) Since flying speeds and routeings vary, aircraft awaiting take-off may not be taken in strict order at departure. On airways under radar control longitudinal separation is normally about 30 n.m. between aircraft, although in the United States this is reduced to 20 n.m. However, since civil radar range is only approximately 200 n.m., most of the world — Atlantic and Pacific oceans, Africa, India, much of Australia, etc. — is without *en route* radar, and in such areas separation is increased to either 10 minutes, 15 minutes, or in certain circumstances to 20 minutes flying time between aircraft, depending on local regulations. At destination, within the terminal control area (TMA), radar separation is reduced to 5 n.m., and when within the vicinity of the airport to 3 n.m. for an approach, giving a landing separation of one

Chicago Control Tower.

minute. (Separation between a smaller aircraft landing behind a big jet may be increased to 6 n.m., once again to avoid wake turbulence.)

On most airways traffic travels in both directions and vertical separation is maintained by the semicircular rule, which allocates specific levels for eastbound and westbound flights *(see* Flight Instruments p.127). On the North Atlantic track system, vertical separation between aircraft on the same track is 2000 feet, but aircraft on adjacent tracks may be at the same level. Longitudinal separation between aircraft on the same track is 10 minutes, and separation between tracks is 60 n.m. Therefore aircraft at the same height, travelling in the same direction on different tracks, have a lateral separation of 60 n.m. In an emergency requiring descent, or a return to base, aircraft fly along the mid position line between tracks.

Air traffic controllers

All aircraft operating within controlled airspace come under the direct supervision of air traffic control officers at all times. Control begins at the departure airport with the ground movements controller (referred to as 'ground') who, together with the take-off and landing controller (referred to as 'tower'), sits high in the control tower with a grandstand view of the runways and surrounding airport areas. At airports with parallel runways, two 'tower' controllers operate on different frequencies, one controlling take-off traffic on one runway, the other controlling landing traffic on the parallel.

Tower Controller monitoring landing aircraft.

Located within a dimly lit room, situated below the airport controllers at the top of the tower, are the approach controllers (referred to as 'approach'), who control landing aircraft within the vicinity of an airport using radar with a maximum range of 50 n.m. Departure controllers (referred to as 'departure') at some airports share control rooms with approach controllers, and at others are situated in control centres adjacent to the departure airport. Area controllers (referred to by their particular area — Boston, New York, London, Scottish, France, etc.), operate from main air traffic control centres (ATCC).

Departure controllers handle departing traffic within the vicinity of an airport before passing flights on to the first *en route* area controller who also handles airways traffic cruising through his section. As an aircraft proceeds along an airway, passing from section to section through a control area, control passes from one controller to another. Controllers monitor progress by radar, or by position reports from pilots, maintaining contact by radio on a frequency pre-allocated for that particular section of the control area. Within a main ATCC controllers covering a specific area sit side by side in sequence. On leaving a section, a controller gives notice of the change of frequency to the pilot, who simply selects the new frequency and re-establishes contact with control by calling the next controller in line. On reaching the control limits of an ATCC, control passes to the adjacent centre *en route,* and the process repeats itself from section to section, and from centre to centre, throughout the flight. Contact between main air traffic control centres is maintained by ground telephone link.

The flight plan

For all flight within controlled airspace a flight plan must be filed with ATC at the departure airport a certain time before start up. The flight plan contains details of the departure, destination, and diversion airports, the route and requested flight level, as well as the aircraft call sign, registration and type. The estimated time of departure (ETD — usually the scheduled departure time) is also given, together with estimates for certain points along route based on the ETD, and such other ancillary information as true air speed (TAS) and Mach number *(see* Flight Instruments p.125), Selcal code *(see* Radio p.59), and radio equipment carried. The flight plan details are then telexed to ATCC along the route.

The flight plan is, of course, only a *request* for a particular routeing and flight level, although with scheduled services approval is normally routine. However, at some time before take-off the pilot requires an ATC clearance indicating acceptance of the flight plan, or, if unacceptable because of conflicting traffic, a re-clearance may be required on a different routeing. The time at which an ATC clearance is received by the pilot varies from country to country, but in any case can be given by ATC only close to departure time to allow co-ordination of the flight with other traffic. Once departed the actual take-off time of the aircraft is telexed to *en route* ATCC and estimates for reporting points are updated accordingly.

In practice most big jet flights are scheduled services operating on a regular basis, and published departure times and routeings seldom vary. The inevitable departure delays do of course occur, and, *en route,* unexpected winds can mean aircraft arriving at reporting points early or late, upsetting flight plan estimates. Therefore, in spite of detailed flight planning information being received in advance by centres, aircraft are handled by each controller as they arrive, and re-routeings or flight level changes may be necessary to maintain separation. In the congested airspace of Europe and the United States, aircraft proceed along a web of airways on different routeings, at different heights and speeds, and in opposite directions along the same airway. Intersecting airways may mean aircraft being required to speed up or slow down, or be vectored by radar (i.e. given headings to steer) to maintain separation. Aircraft departing heavy may be climbing slowly for some time through the levels of congested airways before reaching cruise altitude, while others require descent clearances to destination. Meanwhile, amid this mixed mesh of fast flowing aircraft, adequate separation between flights has to be maintained at all times, a task only possible in the more congested areas of the world with computers and radar, and highly skilled controllers.

Air traffic control clearances

At the departure airport the initial ATC clearance contains the standard instrument departure (SID) routeing, which maintains aircraft clear of noise sensitive areas (The SID varies according to the runway in use and the aircraft flight routeing.) Usually 'cleared to destination, flight plan route' is included, indicating the routeing has been accepted as filed. In theory the initial clearance only covers that section of the *en route* airways within the country of departure, and the pilot is required to obtain re-clearance *en route* when crossing country borders, although flights between friendly states normally proceed without such protocol. However, in a number of areas of the world, (e.g. Eastern Europe, Eastern Mediterranean, Near and Far East, Arabia, Africa, Asia, etc.), certain countries are extremely sensitive about their airspace and care has to be taken to obtain onward clearances before entering.

In many areas civil ATCC are still without *en route* radar, and controllers can only monitor aircraft progress by radio position reports. In such areas flight level changes, for example, are still accomplished by aircraft reporting by radio that any conflicting traffic has been seen passing in the opposite direction. If visual sighting is not confirmed by both aircraft, position reports have to indicate that aircraft are well apart before climb clearance can be given. Also, in security conscious regions, although the relevant ATCC has already received flight plan details and knows of the imminent arrival, proper protocol is required to be observed, and re-clearance must be obtained by radio before entering. With many parts of the world on the verge of conflict, such measures are, perhaps, understandable, but often they are time consuming and quite unnecessary.

However, it has not been unknown for fighter interceptors to appear when aircraft enter foreign airspace unannounced, and at least twice in the last decade airliners have been fired at and forced down. Basic interception procedures involve the interceptor fighter positioning in front of the intercepted aircraft and rocking its wings, followed by a slow turn onto course, meaning, quite simply, follow me! Circling overhead an airport and lowering landing gear indicates the intercepted aircraft must land at that airport, and an abrupt break upwards by the fighter indicates the intercepted aircraft may proceed. The interceptor flying alongside the intercepted aircraft and rocking wings means 'comply with instructions'. The intercepted aircraft responds by rocking wings, indicating that instructions are understood and that the pilot will comply. If at night, both flash navigation lights simultaneously with wing rocking. On rare occasions, when open war erupts (as in the Middle East during the last decade) vast regions of airspace are simply closed, involving flights in long detours.

Because of the traffic density, the North Atlantic is another area where re-clearances are required. Flight plans are submitted as a request for a particular track and flight level, but onward clearances are re-issued in turn as aircraft fly *en route* to track entry points. Clearances for westbound flights are received from Shanwick Control (a combination of Shannon and Prestwick Controls) in the UK and for eastbound flights from Gander in Canada. When *en route* to an entry point aircraft establish contact with one of the above controls and restate the requested track and flight level. An arrival estimate for the entry point is also passed and is fed to a computer which calculates track and flight level availability. Meanwhile, the aircraft proceeds to the entry point for the requested track, which is usually allocated for the crossing, although not always at the required flight level. However, with congestion on a requested track an alternative can be allocated, and the aircraft is then required to re-route (normally under radar) to the new entry point before commencing the crossing. On board, the paper work is, of course, for the requested track, and with a resultant re-routeing the First Officer resembles a one armed paperhanger, rewriting all the flight logs by hand for the new track.

Basic air traffic control procedures

Flight plan details of a particular flight are received by all controllers involved from departure to destination, and are condensed by assistants onto flight progress strips as shown in Fig. 9.1. Where computerized radar displays are in use this information is presented on a control screen (described later). The flight progress strip is mounted on a metal backing plate and placed at the top of a slotted rack in turn as control become active. At the departure airport, for example, when 'ground' is first contacted for start-up, the ground controller assumes responsibility for control and places the flight progress strip at the top of 'ground's' slotted rack. Start clearance is given by 'ground', depending on the traffic situation, and may be delayed to smooth traffic flow and ease congestion on airways.

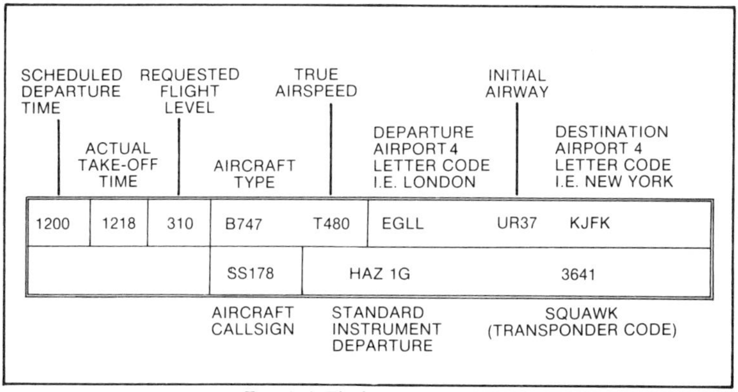

Fig. 9.1 Flight progress slip.

Aircraft departing from European airports bound for the busy routes across the Continent may be issued 'slot times' by which the flight must be airborne, or incur delays. Other aircraft requesting start up in turn have their flight progress strips placed at the top of the rack in a continuous process as each comes under ground control. At some airports ATC clearances are received from 'ground', and at others from a separate 'clearance delivery' frequency ten minutes before taxi. If not, clearance is issued by 'tower'.

'Ground' issues taxi instructions giving directions to the holding point of the active runway, and monitors aircraft movement on the airport from his lofty position in the tower. Taxi guidance lights and runway lighting systems are also controlled by 'ground'. Approaching the holding point of the active runway 'ground' instructs the pilot to contact 'tower'. At this point the flight progress strip is at the bottom of the rack and is removed and passed to 'tower'. 'Tower' now assumes responsibility for control and the flight progress strip is placed at the top of 'tower's' rack. (As the sequence of control develops the flight progress strip mounted on the metal backing can be likened to a relay runner's baton, passing from controller to controller as the flight progresses.) When convenient, the aircraft is cleared for take-off.

Shortly after take-off 'tower' instructs the pilot to change frequency and passes control to 'departure'. Departure control at the base of the tower is passed the flight progress strip by 'tower', in some cases via a tube which runs from top to bottom. When 'departure' is situated at a local control centre departure control is informed by ground telephone of each stage leading to the imminent take-off — aircraft at the holding point, on the runway, taking-off — and so is prepared to receive the flight when airborne. 'Departure' now holds the 'relay baton' and monitors the flight on its instrument departure, issuing radar vectors to steer where required, and clearing the aircraft to climb to higher levels when free of traffic. Outside the airport vicinity, when established on route, area controllers assume responsibility as the aircraft passes from one ATCC to another.

A general view of the Maastricht ATCC Operations Room. (Courtesy Eurocontrol).

At destination, routeing from airways to within the airport vicinity are occasionally along standard terminal arrival routes (STARs), which are now found at a number of airports. During descent, aircraft feed from airways along the STARs, cleared by the last *en route* controller to a final point, usually marked by a beacon, some 20 miles or so from the airport. These points are known as holding points, normally one in each quadrant, and mark the cleared limit of the flight. If landing delays are encountered, aircraft are required to hold over such points, normally flying a four minute race track pattern with right hand turns. These are the so called 'stacks', where aircraft circle one above the other at 1000 feet intervals (from about 7000 feet upwards) awaiting approach clearance. Aircraft descend down the layers of the stack when the lower level is vacated and leave the holding beacon on an assigned heading when cleared from the bottom. Approach control now assumes responsibility and vectors aircraft by radar onto the instrument landing system (ILS). Once established on the ILS, control changes to 'tower' who issues landing clearance. After landing, and safely clear of the runway, 'ground' directs the aircraft to the arrival gate.

Maastricht Air Traffic Control Centre

In the congested areas of the globe, such as Europe and America, the basic ATC procedures previously described have mostly been superseded by modern ATCC using computerized radar systems, although some kind of manual system is normally retained in case of computer failure. One of the most advanced in the world is the Eurocontrol ATCC at Maastricht in the south-east corner of the Netherlands. Callsign 'Maastricht Control', the centre is responsible for upper airspace in Belgium, Luxembourg and north west Germany, one of the busiest areas in the European airways network.

As an example of the system in action, we can follow the progress of an imaginary flight, International World Airways 179 (Callsign, Skyship One Seven Nine) proceeding from London Control, through Maastricht Control, entering at Koksy (Kok) on the Belgian coast, along Upper Green 1 (UG1) in the Maastricht West sector, and Upper Amber 24 (UA24) in the Maastricht East sector, to Diekirch (Dik), near Luxembourg (this route can be followed in Charts p.85). UG1 traverses Maastricht control sector 1A (control frequency 132.2 MHz) and UA24 control sector 3A (control frequency 133.35 MHz).

The Executive Controllers responsible for each sector are aided by Executive Assistants, and backed by Planning Controllers and Flight Data Assistants who program the computers with flight plan details and co-ordinate flights to avoid congestion on the airways. (Because

The operations room, showing on the left the control positions for the Brussels sectors and in the rear the consoles for the Hannover sectors.

(Courtesy Eurocontrol).

Brussels sectors. Planning Control positions showing Display Console and Flight Progress Strip Boards.

(Courtesy Eurocontrol).

of the stressful nature of the work, controllers require a break from the display console about every two hours.) Each Executive Controller, Executive Assistant and Planning Controller has a display console as shown in Fig. 9.2. The circular display screen is not a radar scope, but a computer generated picture that uses processed information received from a number of radar stations. Accuracy is maintained by cross-checking between radar stations, but the display can still function if one radar station fails. Controllers operate the same control sector on a regular basis, and are familiar with airways and reporting points presented on the display screen without identification. Also, most flights are scheduled to depart daily at the same time, and controllers become familiar with estimated arrival times (ETAs) and routes.

The circular display screen indicates transponder-equipped aircraft within the vertical limits of 16,000 to 45,000 feet, but can also be selected to show primary radar returns *(see* Radar p.60). Military aircraft are identified by a star or circle. An individual aircraft is shown on the circular display by a position symbol comprising a small square with a tail of three dots that indicate its previous positions at five second intervals. The tail is known as a 'speed vector'

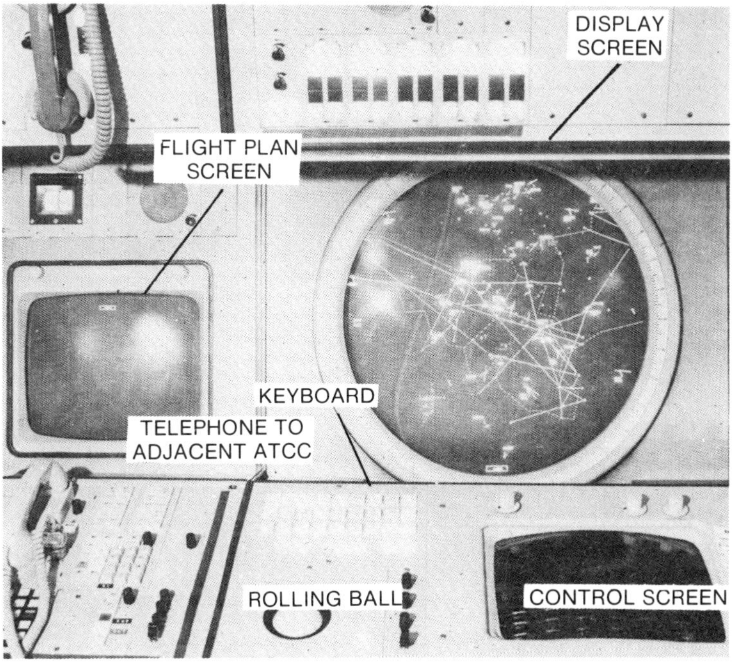

Fig. 9.2 ATC display console. (Courtesy Eurocontrol).

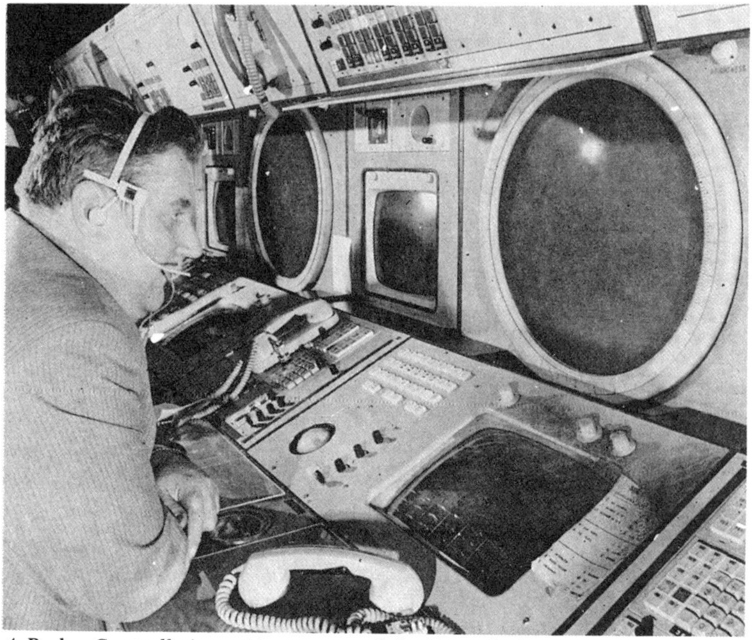

A Radar Controller's position at one of the Brussels Sectors. (Courtesy Eurocontrol).

Control Screen displaying condensed flight plan details as replacement for the Flight Progress Strips. (Courtesy Eurocontrol).

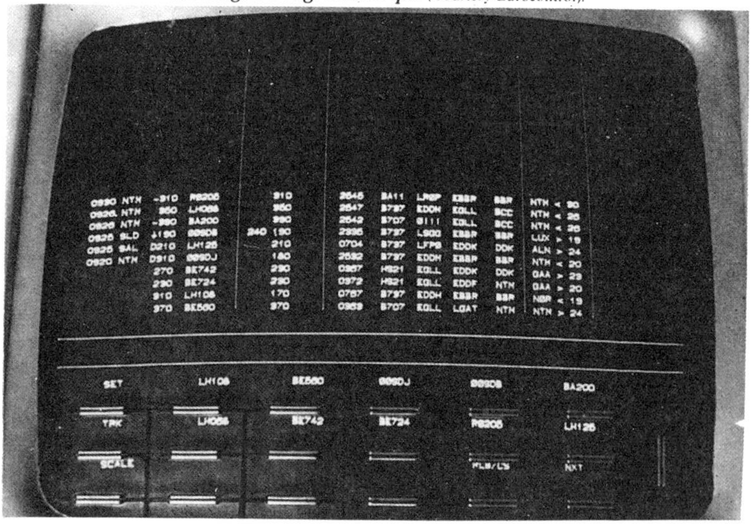

and indicates the direction of travel of the aircraft, the length showing the speed of travel. Alongside the square is a data label showing the callsign of the aircraft and the actual flight level indicated by the transponder. The controller can therefore detect the slightest change of direction or flight level. Each individual position symbol and data label maintains the aircraft position by fading and reappearing further on every five seconds. Aircraft can be positively identified by requesting the pilot to press the 'ident' button on the transponder *(see* Radar p. 61), which results in the position symbol square enlarging and flashing for thirty seconds. Circular display screen movements and radio conversations are recorded continuously and can be played back in the event of an incident.

The control screen below the circular display shows the condensed flight plan details of flights under control within the sector (in a similar manner to the flight progress strips on the slotted rack). Detailed flight plan information of a selected flight can be displayed on the flight plan screen to the left, and includes estimated arrival times at position reporting points along route, as well as standard flight plan information (callsign, departure airport, aircraft type, flight level, etc.). The computer is fed with actual upper wind speeds every six hours and corrects the true air speed (TAS) of flights taken from the flight plan to obtain approximate ground speeds for ETA calculations, updating all times as required.

The rolling ball manoeuvres a sight on the circular display screen in a similar manner to some electronic games, and can be used to obtain track and distance information when vectoring an aircraft to a designated point. Using the rolling ball the sight or curser is first placed over the target and a button pressed. The curser can then be moved to any point and bearing and distance between aircraft and curser is continually displayed.

Telexed flight plan details of Skyship One Seven Nine are received by Maastricht before departure and, if the flight is a scheduled service, will already have been programmed into the computer. If not, the Flight Data Assistant programs the computer with the relevant details. *En route* through the London Control Area, the London ATCC at West Drayton passes to Maastricht an ETA for the boundary between the control areas (a point between Dover and Koksy), the actual flight level of the aircraft (or the level to which it is climbing if the flight has departed London), and the allocated transponder code. These final details are entered into the computer by the Maastricht Flight Data Assistant. If there is any problem in Maastricht accepting Skyship One Seven Nine at a particular flight level the problem can be discussed with London on the direct ground telephone link.

Ten minutes before the boundary estimate, the condensed flight plan details of Skyship One Seven Nine appear at the top of the Maastricht Executive Controller's control screen. As the aircraft falls within transponder range of Maastricht, the computer verifies the aircraft position, before displaying the position symbol on the circular screen, by comparing the flight plan calculated position with the radar position received. Any discrepancies that arise are cleared by the controller. If agreement is detected, the aircraft position is automatically indicated on the display by the position symbol and data label flashing repeatedly until control is assumed. Close to the border, London ATCC instructs Skyship One Seven Nine to call Maastricht on 132.2 MHz.

Once communications are established the Maastricht Executive Controller accepts responsibility for the flight by keying the appropriate selector input button on the computer console. At this point the flashing position symbol remains steady, the condensed flight plan details at the top of the control screen join the list of other flights under control already showing, and at the same time the complete flight plan details, together with estimates for reporting points *en route,* appear on the flight plan screen. Skyship One Seven Nine now proceeds through Maastricht sector 1A monitored by control and maintaining a listening watch on frequency 132.2 MHz.

Ten minutes before the estimate for the change-over point to sector 3A, the condensed flight plan details appear at the top of the control screen of the next Executive Controller in line, operating on frequency 133.35 MHz. Change of control within the control area begins at three minutes before the change-over point, when the position symbol of Skyship One Seven Nine appears flashing on the sector 3A circular display screen. On instruction from the controller on 132.2 MHz, the pilot selects 133.35 MHz and establishes contact with the next controller. Sector 3A controller now assumes control, keying acceptance of the flight into the computer as before. The position symbol and data label once again remain steady, and the flight plan details appear as previously described. On sector 1A controller's display screen, control change is indicated by the position symbol dimming and all flight plan details being erased from the screens. Progress through sector 3A is now monitored by control on 133.35 MHz. At the control boundary, Skyship One Seven Nine is passed to Rhein ATCC at Karlsruhe, in West Germany, and so down the line of ATCCs to destination.

Chapter 10
Flight Crew

Throughout the world flight crew background is wide and varied, and entry into airline flying is often from quite different directions. Many begin flying training in an air force or flying college straight from high school, others after obtaining college or university qualifications, while a few abandon established professions for the lure of flying. The world's air forces are most often the source of pilots entering airlines, but many companies also offer sponsorship on approved courses at flying training schools for selected candidates. Obtaining the basic commercial flying licence on an approved course normally takes one year. A few can afford, or have saved, to pay their own way through flying training school, while many others seek exemption from expensive approved courses by building up flying hours after gaining a private pilot's licence.

Obtaining exemption, however, is not easy, as most governments require applicants to gain extensive flying experience before sitting the required exams and flying tests for the commercial pilots licence (CPL). Many achieve the required experience by first obtaining an assistant instructor's rating and instructing fellow members of a flying club for private licences. Glider towing and parachute dropping can also be a means of obtaining flying hours. Air force personnel also require civil licences before starting airline flying, and many have to obtain them at their own expense.

As well as obtaining a CPL any prospective airline pilot requires an instrument rating (IR), whch involves (for the initial test) a flight in a light twin-engined aircraft in simulated instrument conditions, including holding procedures and instrument approaches. The test is conducted with the examinee wearing a hood to obscure all but the flight instruments. Although all airline pilots are required to retake a similar instrument rating test once per year on current aircraft type (usually in the simulator) to maintain their flying licences current, the initial instrument rating is probably the most exacting flight test undertaken in a pilot's career, because not only is the test difficult but the pilot is relatively inexperienced.

The CPL/IR combination forms the basic requirement for entry into airline service, and all pilots commencing airline flying first have to obtain such licences. As a note, any flying licence entitles holders to fly only aircraft registered in the country in which the licence is obtained. If a pilot wishes to fly aircraft of another country he has first to go through the same procedures to obtain that country's licence unless reciprocal arrangements exist — there is no reciprocal agreement between UK and the USA. However, no matter how the CPL/IR combination is obtained, the road to successful completion is not easy. Selection for air force training or airline sponsorship is rigorous and highly competitive. Many applicants have excellent qualifications, including university degrees, but only very few are eventually accepted. Every attempt is made to select suitable applicants as, in spite of exacting selection procedure, the failure rate on flying courses can be quite high. Flying training is a very expensive business, and costs escalate dramatically in airline or air force service when pilots progress beyond the basic stage. As an example, it has been estimated that the cost of training an air force pilot from scratch to operational standard in a fast-jet squadron is £2.0 million/US $3.0 million.

Air forces and airlines alike therefore, attempt to select the best candidates, but how one rather than another is chosen is difficult to say. Obviously, certain natural qualities such as a degree of co-ordination are desirable, and a variety of aptitude tests are designed to test ability. Of course, almost anyone can learn to fly (or drive, or ride a horse for that matter), but only a few have the innate ability required of the professional pilot (or racing driver or competitive rider) and such tests are designed to seek them out. (In the early days of flying air force pilots were chosen from the cavalry, the assumption being that if one had the co-ordination to ride a horse well, the same standard could be achieved on an aircraft.) Personal qualities such as a certain confidence and self discipline and the ability to be a bit of an individualist who can also work well in a team are desirable — in fact the kind of qualities required in many other walks of life — business, medicine, the Law etc.

The successful few chosen from the many applicants then have to contend with the rigours of the flying course. Those who can privately afford the cost of an approved course may be able to avoid the exacting selection procedures but still cannot guarantee success. A tough year of exams and flying tests lies ahead and if a pass is not obtained, there is no refund. Those who, either by choice or circumstance, obtain exemption from an approved course by building up flying hours the hard way, face big sacrifices to achieve their goal. And, of course, once the CPL/IR is obtained, there is no guarantee of an airline accepting an applicant, and many are required first to find experience elsewhere. Indeed, when recession hits the airline industry, lack of demand even results in airlines refusing to accept those trained under their own sponsorship schemes.

Undoubtedly the most difficult hurdle for the prospective airline pilot is obtaining a CPL/IR and being accepted by an airline (assuming the applicant is medically and academically suited in the first place). Of those who eventually join an airline with the basic licences, many will have only light aircraft experience, as even those with air force training may only have flown fighters, a highly skilled operation itself, but quite different from airline flying. For most applicants, therefore, joining a company is the stage at which airline flying training proper begins. Those starting with the basic licence and only a few hundred hours require about five years of airline flying to complete the training and build up the required experience to have all restrictions lifted. Even during this stage there are those who fall by the wayside and fail to make the grade. On successful completion most pilots are in a position to obtain the highest licence, the airline transport pilots licence (ATPL — UK) or airline transport rating (ATR — USA and Canada), or equivalent, which requires many hundreds of hours' flying experience (depending on the country) and the passing of further exams. Co-pilots with the necessary experience obtain the ATPL/ATR as soon as possible in their career, and all Captains of airline transports are required to hold such a licence.

Once established with an airline transport licence, the road to command for a Co-pilot can, in certain circumstances, be a long one indeed. As Captains retire from an airline, new Captains are promoted (though only after passing a rigorous command course) from the ranks of the Co-pilots in order of seniority, and in some airlines pilots can wait for up to 15 or 20 years from the date of joining before a command becomes available. Most have the necessary experience after about ten years, so the wait can be frustrating for some.

However, whether Captain or Co-pilot, licences have to be maintained throughout one's career, and a number of statutory tests are required regularly — medical (every six months), instrument rating (13 months), aircraft type rating and competence check (six months), safety equipment and procedures check (13 months), route check (13 months), technical questionnaire (13 months), and so on. No other profession is more thoroughly checked. Although it is unusual at this stage for pilots to jeopardize their position by failing technical or flying checks, there is always a fear of losing one's flying

The Flight Crew. The Captain sits on the left. Co-pilot on the right. The Flight Engineer sits slightly aft between the two Pilots.
(Courtesy Boeing Airplane Co.).

licence on medical grounds. At the stroke of a doctor's pen a flying career can be over. However, with constant medical risks, frequent checking, early retirement (55 - 60), a disruptive life and heavy responsibility, the financial rewards can be high, and in a number of European and United States airlines top Captains earn in excess of $150,000 per year.

Crew complement

Big jet crews normally consist of three members, Captain, Co-pilot and Flight Engineer. In most parts of the world (from Europe to Australia and New Zealand) the third crew member tends to be a Flight Engineer (F/E) who does not handle the aircraft but who is a highly trained aircraft engineer. The engineering expertise of such men comes to the fore when engineering problems occur in the more remote airports of the world where trained staff may not be available. Flight Engineers hold aircraft licences that allow them to complete vital work on aircraft which may avoid extensive delays. In the USA and Canada the third crew member on the big jets tends to be a trained pilot who gains his early aircraft experience from the Flight Engineer's seat and who continues his flying training at base to maintain his licence valid. As positions become available he moves to the Co-pilot's seat.

Whoever occupies the third seat, the F/E operates and monitors fuel and engine functions, and manages systems. In airlines where pure (i.e. non-pilot) F/Es are carried, the Co-pilot tends to begin his flying career on smaller aircraft, working his way up to the bigger jets. A pilot begins his airline flying as a Second Officer with one gold stripe on the uniform arm, and gains extra stripes after successful completion of years of airline service; for example, First Officer (F/O) with two stripes after two years, Senior First Officer (S/F/O) with three stripes after eight years, and maintaining three stripes until promoted to Captain when four gold stripes are worn.

The Captain sits in the left hand seat with the Co-pilot on the right. The origins of seating the Captain on the left are lost in time, although the situation has been known to exist since the earliest aircraft with side by side seating. The Flight Engineer sits slightly aft between the two pilots. Flight Navigators are no longer carried on the big jets, navigation now being accomplished by the inertial navigation system. Also, although the masculine gender for crews is used throughout, it in no way implies that women are not to be found on today's flight decks. The number may be few, but the female ranks are increasing all the time, and a handful of airlines in the United States and Europe (United, TWA, Aer Lingus, British Caledonian, UK Air, Dan Air, etc.) already have their share of women pilots. In 1979, the first woman jet Captain, Captain Yvonne Sintes of Dan Air (now retired), was joined on Comets by F/O Marilyn Booth to form the first all-women flight crew. In July, 1984, Captain Lynn Rippelmeyer of People Express became the first woman to captain a Boeing 747 across the Atlantic.

Crew operation

While seated, flight crews must remain strapped in at all times. Even meals are eaten from trays on laps. In the cruise, however, all emergency drills can be completed by only two crew members, and short rest periods are possible. The Captain and Co-pilot normally share the flying turn and turn about, one handling the aircraft on one sector while the other completes the co-piloting tasks of monitoring equipment operation, log keeping, radio work etc., and *vice versa* on the next. Ability among professional pilots varies little, although, of course, some are better than others. However, superior handling ability in an airline pilot is not so important as one might imagine: much of the flight is automatic, and the pilot's job today is essentially one of operations director and systems manager. Pilots tend to talk of a colleague being a good operator rather than a good pilot — anyone who can fly an aircraft with extreme accuracy into the side of a mountain may be a good handling pilot, but a bad operator!

However, pilots *may* have to land large jet aircraft at airports with mountainous terrain using little more than eyesight and the seat of their pants, perhaps at night and in bad weather. It may be that ground equipment is not installed, is unserviceable or under repair, or, because of wind direction an approach to a runway is required over difficult terrain which has prohibited the installation of an instrument landing system. In such cases the pilot employs all the basic flying techniques, perhaps flying a circuit before commencing final approach, and is required to exercise the handling skills of the light aircraft pilot while flying a heavy jumbo jet. Also, as mentioned, autolands are designed for foggy conditions when winds are light and cannot be attempted when cross winds are above 15 knots — or where the air is turbulent, and the aircraft is required to be hand-flown. Such procedures as visual approaches and landings may take only a few minutes, or, like flare judgement on touch-down in a cross wind, only a few seconds to perform, but require much training and practice, and flight crews, like other professionals, can often make difficult tasks look easy. To maintain operating skills and avoid becoming rusty, crew members are required to fly regularly (similar to the practice needed in playing a musical instrument) and indeed, by law a pilot or F/E must fly once every 28 days or is required to be re-checked at base before service.

Crew members may not have met before the flight, and standardization on the flight deck is of the highest order. Even a trained observer would have difficulty in judging whether or not a particular crew had flown together before. (Such a system avoids crews picking up bad habits and avoids disruption to a crew's routine when one member is absent.) Each flight crew member has his own particular duties to perform which form an integral part of the complete operation, and co-ordination between crew members is essential. Flight crews operate very much as a team with the Captain at the head, and in the close environment of the flight deck each carefully monitors the performance of the other two while completing his own tasks. (Most flight decks are quite small and, with the need to have all equipment within reach of the three crew members while seated, seats are placed close together. On the B747 the Flight Engineer's seat can be electrically driven fore and aft to cover the extent of the instrument panel.) Check lists, procedures and drills are ergonomically designed to be efficient and logical in operation, and crew performance is expected to be of the highest professional standard. Crew members do of course make mistakes and omissions do occur, but in the closely monitored environment of the flight deck, where each is used to correcting, and being corrected by the others, the system works extremely well.

On most flights operation is routine with the crew following standardized and practised procedures just like any other professional team in a precise environment — from the operating theatre to the live TV studio — and it is naive to assume that Captains make major decisions every couple of seconds. However, on every flight circumstances change, even when flying repeatedly on the same route and, as in any other practical situation, small difficulties have to be overcome. On the highways, for example, every driver knows the problems of road works, traffic jams, breakdowns, diversions, and weather, etc. In flying, work at airports, take-off and landing delays, malfunctions, re-routeings, weather and so on, also present similar problems. Like other professionals, however, crews not only have to perform the routine well, a highly skilled procedure in itself, but also are trained to cope with any emergency. On the rare occasion when real emergencies do arise — severe weather, system malfunction, engine failure, aircraft fire, etc. — decisions are made, sometimes in a split second, which can affect the safety of the aircraft and perhaps many hundreds of lives. Here the Captain comes into his own, and the training of the crew is put to the test.

Crews have a little knowledge of many subjects and, as Captain Ian Frow of British Airways says, 'It's a matter of being a jack of all trades to be master of one!' Details of the flight (weather forecast, fuel, flight plan, loading, etc.) have first to be thoroughly checked before departure. Although much of the paper work is completed by ground staff, the practical operation of the aircraft may not always be as manuals, graphs and statistics indicate, and crews have the specialized knowledge to effect any last minute changes that may be required, the final decision being made by the Captain. The crew is also the last line of defence against any errors being carried into the air. In the final analysis it's the crew who 'carry the can', and the ultimate responsibility rests firmly with the Captain. The buck definitely stops there!

Crew operational practices are difficult to describe and drawing analogies even more so. Obviously take-off and climb, descent and landing, are the busiest phases, but even here crews can be observed in little physical activity, the flying of the aircraft mostly being accomplished by the autopilot. Pilots in this situation are frequently likened to concentrating chess players, each often physically immobile, but with minds active. Like the chess game, big jet aircraft operation consists of a number of basically simple moves that combine to form a complex whole. Crews can hardly be described as the equivalent of Grand Masters, but the basic big jet flying moves run into many thousands, and being caught in 'check mate' is definitely not allowed. Add to this the responsibility for many hundreds of lives, plus the high financial value of aircraft and contents (a Boeing 747 costs approximately $100M), and it is not a simple game to play.

In spite of the sophisticated equipment, crews are trained to be sceptical of automatics. The more reliable the apparatus the more easily crew members can be caught off-guard when malfunctions occur. Keeping ahead of a fast jet aircraft requires concentration and alertness, especially during take-off and landing, and although the autopilot may be in control, each pilot continuously updates a mental picture of the aircraft position and operational requirements throughout the flight. It may be night, or the aircraft in cloud, and, without the advantage of electronic signals, the brain is required to build a mental picture of the overall scene by observing instruments, listening to other aircraft's radio reports, tuning radio beacons, etc., for the monitoring and cross checking of automatics. It is a bit like viewing the solar system from earth. In fact for clarity, models normally demonstrate the solar system by imagining the observer viewing the system from the outside. When observing aircraft traffic from an airport perimeter the picture is similarly quite clear. However, when trying to construct a mental picture from the inside, on board an aircraft and using only one's basic faculties, it's not quite so easy. It is not surprising that the early astronomers had the sun going round the earth!

The autopilot, too, requires attention, as it can do no thinking for itself nor listen to air traffic control instructions, and pilots have continually to feed the required information to the automatics. If wrong details are inserted the autopilot obeys without question, so vigilance is of the highest importance. Basic airmanship (i.e. the collective practical application of training, skill, experience and professional judgement) is required to be exercised by all flight crews at all times.

Crew conditions

Flight crew conditions vary according to airline, aircraft type, and route network. Some crews operate continuously on a fixed route while others fly worldwide wherever their aircraft type takes them. In some airlines crews are simply allocated trips on a roster while in others a bid line system operates, in which individual crew members can bid for available work like bidding for a lot at a postal auction, the more senior the crew member the better the chance of making a successful bid. Many short haul airlines operate on a six-days-on, three-days-off routine with crews at home most nights, apart from the odd night stop away. On long haul flights crews can be away for up to two weeks at a time; perhaps 180 days, or even more, away from home in any one year.

While away from base crews are accommodated in first class hotels, and allowances for food are paid separately in either local currency, US dollars (or equivalent) or travellers' cheques, depending on airline policy. Short haul crews may only have the occasional 24-hour stop to visit foreign cities, while long haul crews usually have rest periods down the route. However, since most flights are on a daily basis, crews are generally kept on the move. A New York-based crew, for example, on a round the world flight operating New York, Anchorage, Tokyo, Hong Kong, Delhi, Bahrain, London, New York, on a daily basis, could in theory be round the world in ten days, working up to ten hours on duty each day, with a one day stop-over in each place. This allows for an eight-hour flight followed by a twentyfour-hour period at the transit stop, with a fresh crew taking the aircraft on to the next destination. (During transit, although time changes may be quite large, most crew members attempt to live by local time.) Twenty-four hours later the resting crew operate the next service through, and the process is repeated.

Ideally 12 or 36 hours are the best crew rest periods, allowing the crew to recommence duty after one or two sleep patterns respectively. Commercial requirements, however, usually result in a 24-hour rest period. Since the normal 24-hour daily rhythm comprises eight hours sleep followed by 16 hours awake, crews often return to duty after 24 hours' rest with only one sleep period and a second about to begin!

North Atlantic trips tend to be westbound during the day and

Chocks positioned by the nose wheel to prevent the aircraft rolling. Crew flying times are recorded from chocks away to chocks under.

eastbound at night. European crews, therefore, operate to North America in daylight, spend one night and part of the next day at destination, and then operate back overnight. American crews operate eastbound at night, spend one day and night in Europe and operate back in daylight the next day. By such means aircraft are kept more or less continuously on the move by a constant supply of fresh crews. On the other hand, an airline with a weekly flight to a far off exotic island, like Mauritius in the Indian Ocean, could have a crew waiting in the sun for a week while the next service comes through (more like a free vacation), while some airlines base crews with their families in foreign cities for several months at a time.

The over-riding restriction for flight crew is actual flying hours which is generally set by law at a maximum of 100 hours in any 28-day period. By law crews must note all flying hours in a personal log book, the times in GMT being taken from 'chocks away' to 'chocks under'. Critics say crews are paid a lot of money for doing a part-time job, but although 100 hours maximum per 28 days doesn't sound much, crews can in fact be on duty for quite considerable periods while building up only a few actual flying hours. On a short half-hour flight, for example, the crew may complete three return journeys in one day totalling three hours but, with flight preparation and waiting time on the ground at either end, could be on duty for eight to ten hours. On a long haul operation, flying say Colombo-Seychelles-Johannesburg, the total flight time is around nine hours, but the complete duty day is nearer 12 hours, and might even begin at midnight. Indeed, with many long haul flights transiting stations at all times of the day and night, much work is at night (e.g. a long distance Europe - Australia flight is arranged to depart and arrive at reasonable times but will often transit stations *en route* in the middle of the night). On rare occasions, delays can result in crews being on duty for up to 16 hours (and more regularly up to 13), which is equivalent to working in the office from 9.00 a.m. until 1.00 a.m. followed by a difficult drive home.

To help prevent fatigue, crew scheduling limits restrict by law the maximum duty day that crews can operate, depending on the time of departure, the number of sectors to be flown, and whether commencing flight at home or away from base. However, in spite of regulations, prevention of fatigue can be a problem, and crew members react differently in different circumstances. Like every other traveller the flight crew has to cope with all the usual tiring effects of travel, as well as fly the aircraft. For administrative purposes, however, crews are listed on a general declaration, and are normally processed through immigration and customs separately, which undoubtedly eases the strain. (In almost every country customs regulations for the crew are more restrictive than those for passengers.) Fatigue is often the result of a number of factors, and on long haul flights can be exacerbated by low moisture content and low oxygen intake, due to cabin altitude. The effect of large time changes, extremes of temperature and weather, exposure to alien food and water (leaving Calcutta, happiness is a dry fart), changing biorhythm (any complex task highlights even the slightest variation in performance), or perhaps even something as simple as being unable to sleep in the hotel during the day before a night departure (every hotel seems to employ a duty hammerer) can all take their toll. As regards drinking and flying, crews are not allowed to drink alcohol from between eight to 24 hours before commencing duty, depending on the laws of the country (e.g. UK, eight hours; USA, 24 hours).

Part 2

The Flight

Chapter 11
London to New York

It is assumed that readers are familiar with the preceding chapters so 'The Flight' is written mainly in aviation terms. Although most flights are routine, no two journeys are ever the same, and this account is of the progress of a typical flight.

Flight planning

In operations, paper work for the flight is spread across the counter. Operations staff have already prepared details for checking and the crew examine the information presented before acceptance. It is now less than one hour to departure and a busy period lies ahead. At exactly one hour before departure the crew signed on, checked the latest company notices, mail, etc., and if at base and they had not already met would have taken the opportunity to introduce themselves. On Duty time begins with signing on.

The crew of Captain (Capt.), Co-pilot, of First Officer (F/O) rank, and Flight Engineer (F/E), are operating International World Airways westbound round the world service (callsign Skyship One) on today's sector from London to New York. Departure time is 1400 G.M.T. (the same as London local time in winter) with a scheduled arrival time of 2135 G.M.T. in winter, local arrival time 1635, i.e. New York is 5 hours behind G.M.T. in winter, giving a scheduled sector time of 7 hours 35 minutes. The sector time includes taxying, take-off and landing delays, etc., and the actual flying time from airborne to touch-down of today's flight is 6 hours and 57 minutes.

The longest scheduled non-stop service is operated by a Pan Am Boeing 747 SP from Sydney to San Francisco with a sector time of 13 hours 25 minutes. The longest delivery flight, also by a Boeing 747 SP, was from Seattle to Capetown (8936 n.m.), a total time of 17 hours 22 minutes. Two hours of fuel still remained on landing; an amazing achievement in spite of the aircraft being flown empty. However, in spite of bigger aircraft with larger fuel capacities and sophisticated electronics, conditions are not necessarily made any easier for crews in normal operating conditions. Costs influence a Captain to carry the minimum of fuel commensurate with safety and at times economic considerations require aircraft to be flown to design limits, resulting in big aircraft carrying heavy loads over very long distances with landings in severely reduced visibility.

It is now mid-February, approaching the end of winter in the USA, and the Captain is carefully examining the New York weather. The coded weather report (Fig. 11.1) indicates the weather from 1200 hours G.M.T. on the day of departure for a 24-hour period until 1200 hours G.M.T. on the following day to be: wind at 060°T at 11 knots, visibility 2400 metres, rain, cloud cover 8/8ths stratus at 300 feet; intermittently between 1200 and 2000 G.M.T. visibility 4800 metres, rain, 8/8ths stratus at 700 feet; gradually between 1900 and 2000 G.M.T., wind 250°T at 14 knots, visibility more than 10 kilometres, weather nil, 8/8ths stratocumulus at 2000 feet; between 2000 and 2300 G.M.T., 40 per cent probability of weather becoming 3200 metres, rain and snow, cloud 8/8ths stratus at 1000 feet; gradually between 2200 and 2300 G.M.T., wind 290°T at 17 knots, gusting 28 knots, cloud 5/8ths stratocumulus at 3500 feet; gradually between 0500 and 0700 G.M.T., cloud and visibility OK; gradually, 1000 to 1100 G.M.T., wind 270°T at 15 knots.

```
NEW YORK ( JFK ) CODED TERMINAL AREA WEATHER FORECAST

TAF
KJFK 1212 06011 2400 61RA 8ST003 INTER 1220 4800 61RA 8ST007
GRADU 1920 25014 9999 WX NIL 8SC020 PROB40 2023 3200 83RASN
8ST010 GRADU 2223 29017/28 5SC035 GRADU 0507 CAVOK GRADU 1011
27015=
```

Fig. 11.1 New York coded weather forecast.

The forecast indicates the weather improving at arrival time, but still not very good. Cloud cover extends over most of the eastern seaboard, and all major cities are affected. Boston, normally the diversion airport, is forecasting 8/8ths nimbostratus at 500 feet, with visibility one nautical mile in snow, which is below the diversion airport minimum limits of 800 feet cloud base and two statue miles. Snow is common in the USA in winter and severe snow storms can close all major airports on the eastern seaboard within a matter of hours. If the weather does not clear as quickly as forecast it could be bad in New York at the time of arrival, resulting in delays, and since only enough fuel is carried for flight requirements it is imperative that sufficient is on board.

An inspection of Canadian airports shows Montreal forecasting a wind of 270°T gusting 15 to 20 knots, visibility more than 10 kilometres, 8/8ths stratocumulus at 3000 feet. Montreal is nominated by the Captain as diversion airport with a decision to carry full contingency fuel in case of landing delays at New York. The maritime airports of Canada (Gander, Halifax, St. Johns, etc.) are forecasting low cloud with heavy rain and strong winds. Gander seems the most likely in the case of *en route* diversion. London and Shannon forecasts are also checked and indicate the expectation of cloud and rain, but with good visibility. Both are suitable return alternates if problems arise at the beginning of the flight. Upper air charts (similar to those in the Met. section) are also inspected for upper winds, forecast CAT, and any significant weather expected *en route.*

```
ROUTE 05/100 FL310  53N015W/FL350
MINIMUM COST PLAN  -  REQUIRED FUEL 102174 KG
                      ONE ZERO TWO ONE SEVEN FOUR

STD  1400  STA  2135  TOT  0735

-B747/H-S/C
-EGLL1420
-0499F310 UG1 STU UB10 CRK 53N015W/M084F350 NATF SG/0490F350
 NA140 ENE DCT ENE219085 V16
-KJFK2117 CYMX
-REG/XTFOF SEL/ABCD
EST/15W1538 SG/1904

GROUND DIST 3076  AV W/C M036
TOW 304874 KG          ZFW 203700 KG             LWT 224700 KG

TRIP FUEL             80174 KG     6.57      PLD REM 21000 KG
DIVERSION             12116 KG       45      YMX
CONTINGENCY            4500 KG       30
RESERVE                4384 KG       30
REQUIRED FUEL        101174 KG     8.42
EXCESS                    0 KG
TAXI                   1000 KG
TANKS                102174 KG    -OPERATIONAL PLAN NO.  1

DIVERSIONS CYMX FL330 P 27 430NM 12116 KG  T45 DIV FUEL
           KBOS FL170 P  8 204NM  8022 KG  T34

EQUAL TIME POINT  3.41
RMKS

A/C PERF DEFECTS- NONE
```

Fig. 11.2 Air Traffic control and fuel flight plan.

An examination of the flight notices shows nothing of note that may prevent departure (runways or airports closed, airspace restrictions, etc.). The usual work in progress is noted at London, some minor light systems and one radio frequency unserviceable at New York, and nil for Montreal. A military exercise is in progress in the USA with a number of airways closed, but none affecting the flight. A final check is made of the completed fuel flight plan figures by a simple rule of thumb, and the calculations scanned for error. The total required fuel figure of 102.0 tons comprises 80.0 tons fuel for London, Heathrow (LHR) to New York, J.F. Kennedy (JFK), 16.5 tons diversion fuel (including reserve) to Montreal (just in case), contingency fuel of 4.5 tons, and 1.0 ton of fuel for taxi at London (Fig. 11.2).

Flight logs for the complete journey (Fig. 11.3) display routeings

FLIGHT LOG

POSITION IDEN/FREQ MSA AWY	NOTE TRM	 DRF	LAT LONG TRT	PFL TMP WIND	MNO GS DIS	 ETA TIM	 RETA	 ATA TTLT	GDTG AWTG	FOB REQ
LONDON			N51 292						3076	....
LON113.6			W000 279			....	....	....	M036	802
2.4 SID	-268-	01L	262.6	32008	16	6		6		
-WOODLEY			N51 272	100					3060	
WOD 357.0			W000 527	P03	360	....	....	....	M036	
3.1 SID	-285-	01L	278.2	32009	42	8		14		
-ABM LYNEHAM			N51 330	210					3018	
			W001 599	M18	358	....	....	....	M037	
3.1 SID	-281-	02L	274.8	34017	12	3		17		
-HTN			N51 340	245					3006	
116.3 BCN	112R36D		W002 198	M25	356	....	....	....	M037	
4.0 SID	-292-	04L	285.6	34026	36	6		23		
-BRECON			N51 434	285					2970	
BCN116.3			W003 157	M33	351	....	....	....	M038	
4.2 UG1	-292-	05L	284.5	34036	66	6		29		
-TOC				310					2904	....
				M38	344	....	....	....	M038	707
4.2 UG1	-292-	04L	284.5	34037	2	0		29		

Fig. 11.3 Flight Log.

and are required to be checked against the ATC flight plan routeing (Fig. 11.2), which is filed with ATC by operations staff before departure. The requested routeing is Upper Green One to Strumble, Upper Blue Ten to Cork, at flight level 310, then track 'Foxtrot' from Cork, at flight level 350, on co-ordinates 53N 15W, 54N 20W, 55N 30W, 54N 40W, 51N 50W to Springdale on the Newfoundland coast of Canada. From Springdale, North American route 140 to Kennebunk just north of New York, and on to JFK.

The North Atlantic track system consists of about six approximately parallel tracks active from entry points on the coasts of the UK and Ireland to exit points on the Canadian coast, and, of course, *vice versa.* (Normally westbound flights by day, eastbound by night.) The tracks are selected twice daily by computer to avoid the strong head winds when flying from Europe to the USA (designated tracks A, B, C, D, E, and F), and to follow the best winds when travelling in the opposite direction (designated tracks U, V. W, X, Y, and Z). The requested track depends on departure and destination of a flight, and for Skyship One track 'F' gives the quickest flight time. Final track allocation depends on availability of space on the desired track.

North Atlantic track details and co-ordinates are also thoroughly cross-checked for accuracy. With so many aircraft crossing in the same direction at the same time on roughly parallel tracks, a simple error in one co-ordinate could create conflict. Airlines tend to schedule departures to similar destinations at much the same time. Since all big jet aircraft prefer to operate at flight levels within the narrow 11,000-feet band from 28,000 to 39,000 feet congestion often results, especially over such busy areas as the North Atlantic.

With the checking of the paper work complete, the crew board transport to take them to the aircraft, registration X-TFOF positioned at stand kilo 14 on the west side of the airport. (The first aircraft registration letter designates the country, e.g. D- West Germany, F- France, G- UK, N- USA, etc., with the remaining four similar to a vehicle registration. Aircraft are normally referred to by the last two letters, in this case 'oscar foxtrot'.) Kilo 14 is a stand where aircraft park nose-in and where passengers board the aircraft by the forward doors via a covered walkway, which extends like a finger from the terminal building. It is now approximately 40 minutes to departure and many checks and procedures are yet to be completed.

Pre-flight

On board, the Captain and Co-pilot commence the pre-flight checks while the Flight Engineer completes his walkround inspection to check for unserviceable equipment, damage and leaks, and to examine control surfaces, skin condition, and wheels and tyres, etc. (Tyres are not changed on a regular basis, but only when required. Where tyres are badly scuffed or have cuts of more than a certain depth they must be changed, but tyres slightly worn can be changed at a more suitable time). A maintenance log is normally first examined by the Captain to check aircraft condition, and any relevant defects that may affect performance are noted. A list of acceptable defects, such as minor malfunctions, which can be deferred for maintenance and with which a flight can depart, is contained in a despatch deviation manual.

Hangar Maintenance.

Aircraft maintenance, of course, is a continuous and on-going concern, and at every turn-round and transit station routine checks are conducted, any minor faults being cleared, if possible, in the time available. Often the cause is nothing more than dust, moisture or slightly loose connections, and removing and re-racking equipment is frequently sufficient to clear the fault. Any uncleared faults are listed in the maintenance log and referred back to base. Where unserviceable equipment is permitted to be carried flight crews may be inconvenienced, but the situation is not considered unsafe. More serious malfunctions (main flight instrument failures, flying control or engine problems, or landing gear or flap malfunctions, etc.) must be rectified before flight, and often result in delays. Although annoying for passengers and crew, safety is the overriding factor.

At base, hangar maintenance is conducted according to manufacturers' schedules, which stipulate the frequency and degree of inspection and equipment replacement, depending on accumulated flying hours, numbers of landings, etc. Preventive maintenance is the order of the day, with aircraft being serviced at regular intervals. Condition monitoring is now also standard practice, in which inspection procedures and in-built monitoring systems give a regular and up-to-date indication of equipment condition. All Boeing 747s, for example, are fitted with airborne integrated data systems (AIDS) which record and store basic flying, engine, autopilot and system parameters throughout each phase of the flight. Engine and equipment performance can therefore be monitored at regular intervals and adverse trends detected at an early stage. In addition, engine health is regularly checked by completion of the Flight Engineer's 'engine condition monitoring log' on every sector over three hours.

A program of metal fatigue and corrosion analysis is also conducted during hangar checks. Today, aircraft are built with lives of at least 25 years, and are quite strong enough to cope with the daily stresses of operational flying. Extremes of weather, large temperature changes (35°C on the ground to -60°C in the cruise is not uncommon), wing flexing, landing and take-off loads, and engine vibration pose few problems. Individual aircraft, however, may experience stressful incidents such as heavy landings, bad weather or severe turbulence, and spillage from dangerous cargoes, toilets or galleys may cause corrosion. The detection of corrosion or cracks indicating metal fatigue therefore forms an important part of every service. In the hangar aircraft sections can be X-rayed, and deep analysis of vital areas can be conducted by the use of ultra-sonic crack detection techniques. Localized cracking (in the vicinity of fasteners, etc.) can be detected by apparatus measuring the distortion of eddy current waves. In spite of such modern equipment, however, visual inspection still plays a vital role, and nothing can replace the old fashioned white-coated inspector searching with torch and magnifying glass.

On the flight deck now, the Co-pilot switches the three INS sets to align, and after checking the equipment inserts in each the present position, noted from a manual (i.e. LHR 51,28.3N; 000,27.5 W) by keying the appropriate numbers. Since the alignment process takes about 13 minutes and the aircraft cannot be moved until alignment is complete and the INS selected to navigation mode, it is prudent to switch on the system early. Also, instrument checks cannot begin until the INS gyros spin up and the displayed failure flags retract.

The relevant paper work is then distributed among the crew — a copy of the flight log to each, fuel flight plan to the Flight Engineer and such papers as the ATC flight plan and flight notices, etc. to centre console positions, available for easy reference. A large number of books, manuals and documents are carried on each aircraft (referred to as the library) and are checked by the Co-pilot against a list of requirements for the area of operation.

On retraction of the instrument failure flags, the pre-flight instrument and equipment checks can begin. The procedure is known as a 'scan check' and each and every item is checked in sequence from memory. The scan commences at top right on the pilot's overhead panel, flows up and down through each piece of equipment, down across the autopilot and autothrottle switching on the glare shield, zigzags across the pilot's centre panel and the Co-pilot's instrument panel (omitting the left instrument panel for checking by the Captain) and finally along the various items on the centre console. The checks are too numerous to mention individually but include oxygen check, radio switching, cockpit voice recorder test, warning lights test, ground proximity warning test, flight instrument checks, take-off configuration warning check, trim settings, and window heat switching and check (flight deck windows are five to six centimetres thick and are heated to clear frosting, and to prevent cold soaking in the low temperatures, causing brittleness. Window strength against bird strikes is tested in the factory by firing chickens at the laminated sections from a special cannon!).

Once aboard, the Flight Engineer begins a similar scan check relevant to his instrument panel, checking and setting cabin altitude controls, checking fuel systems and monitoring refuelling, checking fire warning indications, engine instrument displays and safety equipment, etc. He also examines the maintenance log and confers with the Captain and First Officer on any defects affecting performance.

Meanwhile the Captain completes his pre-flight instrument check and begins to load the INS with the positions of the first nine way-points along route, while the Co-pilot commences the take-off calculations. The current weather conditions are first copied from the ATIS, which at the moment is information 'sierra' — wind 250°M at 15 knots, temperature 12°C/dew point 8°C, altimeter 1023 mb, departure runway 28 left. From the take-off manual the maximum permitted take-off weight in today's conditions is extracted from a graph constructed from the details of runway 28L at Heathrow, and the weight noted on a take-off proforma. The graph is constructed with runway length and gradient in mind, and is entered with wind component and temperature. The maximum take-off weight indicated provides sufficient obstacle clearance with one engine out for the first segments of the flight, i.e. initial climb, gear retraction, flap retraction and *en route* climb. (Maximum take-off power can be maintained on three good engines in the event of failure for up to ten minutes after departure. With an engine failure on take-off, flap retraction is commenced at 1000 feet above the departure airport.)

Once the actual take-off weight is known the proforma can be completed with take-off speeds and engine power setting. Take-off weight is expected to be in excess of 300 tons, but the exact weight is not known until the loading is complete and the load sheet is presented to the Captain for checking and signature about ten minutes before departure. The stabilizer trim setting is also obtained from the load sheet. The Flight Engineer and First Officer cross check from engine charts the maximum engine pressure ratio (EPR) available (i.e. 1.44 EPR) and this figure is also noted on the take-off proforma. Since the take-off weight is less than maximum, something less than maximum take-off power is sufficient, and a graduated power take-off can be considered.

Calculation of the take-off speeds of V1, V2 and VR (V stands for velocity) is as follows. V1 is the go or no-go decision speed. In the event of an emergency occurring before V1 sufficient runway is available for stopping, but after V1, the aircraft is committed to take-off. In fact, V1 is only critical at heavy take-off weights where abandoning at high speed and high weight is a hazardous procedure. Overheating brakes and bursting tyres are likely and if swift action is not taken, over-running the runway length is a possibility. Since most take-offs are at something less than maximum weight, V1 becomes less critical with reducing weight. It may seem sensible to maintain V1 at the same speed for all take-offs, but, in fact, at weights less than about 290 tons the lift-off speed (i.e. rotation speed, VR) on many runways is less than the V1 speed at maximum take-off weight. Since a V1 speed greater than rotation speed is nonsense (you can't exactly decide to abandon once the aircraft is airborne), V1 is arranged to be a set number of knots (usually 10 - 20 knots) below VR for a particular take-off weight. This however has the anomalous effect of the decision speed, V1, *increasing* with increasing take-off weight, when one would imagine it to decrease.

V1, in fact, is the maximum speed at which an emergency requiring the take-off to be abandoned can be recognized, so while action is being taken the aircraft may continue to accelerate a few knots past V1 before deceleration commences. To all intents and purposes, therefore, the speed at which an aircraft commences deceleration on an abandoned take-off, and rotation speed, are very close and V1 and VR become critical only at maximum take-off weights on restricted runway lengths (Fig. 11.4). The number of knots reduction from VR to obtain V1 is indicated on the graph page of 28L in the take-off manual. On a wet runway a further reduction from VR in the region of 20 knots is required to obtain a 'wet V1'. On some wet runways the danger of aquaplaning is a possibility — traction is impaired at speed due to wedges of water building up between the tyres and runway surface. Standing water, slush or snow on the runway are only

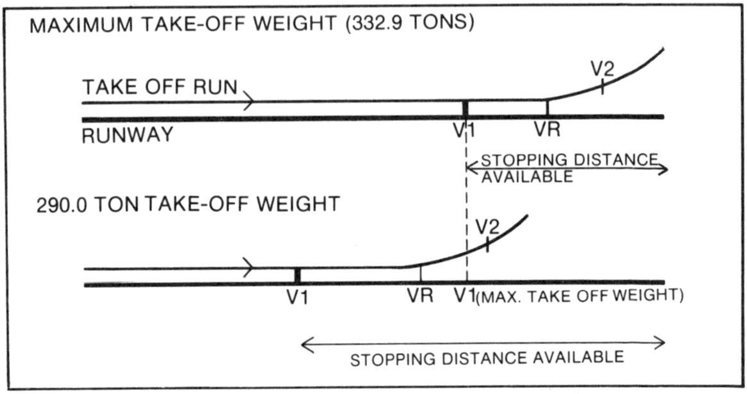

Fig. 11.4 Effect of take-off weight on stopping distance.

acceptable up to certain maximum thicknesses, above which take-offs are not permitted. Where take-offs are possible in such conditions severe restrictions on the take-off weight are imposed which effectively reduce the take-off speeds. A reduction in the take-off weight may also be required for defects such as an engine producing slightly less than full power — known as a 'shortfall' — or more than two brake anti-skid channels being inoperative, etc.

The lowest acceptable V1 speed is the minimum control speed on the ground (VMCG), which is around 122 knots at sea level. Below VMCG there is insufficent airflow across the rudder surfaces to maintain the aircraft straight with an outboard engine failure at take-off power setting if the take-off were to continue.

The take-off speeds of VR and V2 are extracted from a table against actual take-off weight of the aircraft. VR, the rotation speed, is the take-off speed for that particular weight at which the pilot 'rotates' the nose of the aircraft to the nose up attitude required for lift-off, and V2 is the safe climb-out speed required in the event of losing an engine at V1. Normal climb-out speed is V2 + 10 knots. The minimum speeds for retracting flaps are also based on V2 and are as follows: minimum speed for retracting flap from 10° to 5°, V2 + 40; from 5° to 1°, V2 + 60; and from 1° to flaps fully in, V2 + 80.

At this stage the Captain is examining the instrument departure procedures, tuning beacons, and setting VOR radials in the course window for the required routeing. The departure clearance is not issued until engine start is requested, but the runway in use is known

(HEATHROW) **LONDON**

BRECON **SID's**

Trans alt **6000**	G2
1. Initial climb: Ahead to 500(QFE) before commencing turn. En-route cruising level will be given by London Control after take-off. 2. Communications: Do not change from TWR to SID freq until advised. Callsign for SID freqs marked * will be 'HEATHROW Approach'. 3. Max 250kt IAS below FL100 unless otherwise authorised. 4. Noise Measurement Sites, see C1.	**EFF 22 DEC 83**

EGLL

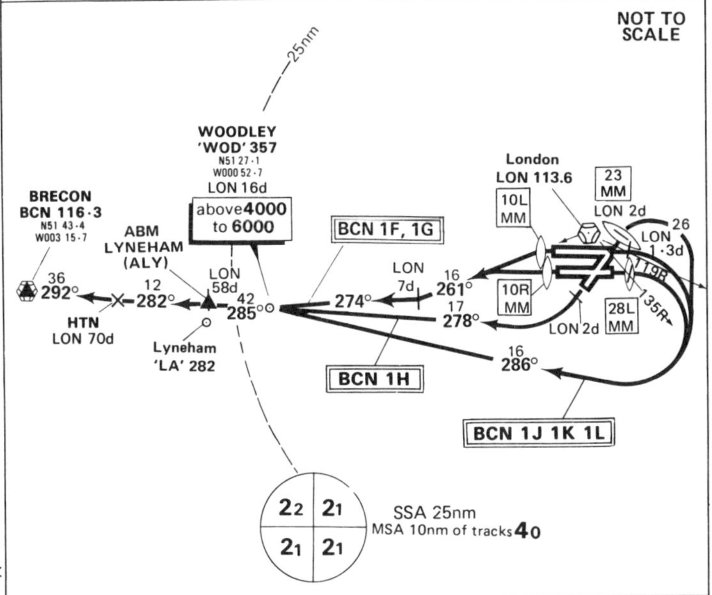

A

SID	R/W	ROUTEING (including Min Noise Routeing)	ALTITUDES
BCN 1F 132.05	28R	Ahead to intercept LON 261R. At LON 7d right onto Tr 274M to 'WOD'. At 'WOD' right on Tr 285M to ALY then Tr 282M to HTN then right on Tr 292M to BCN.	'WOD' above 4000 To 6000
BCN 1G 132.05	28L	Ahead to intercept LON 261R. At LON 7d right onto Tr 274M to 'WOD'. At 'WOD' right on Tr 285M to ALY then Tr 282M to HTN then right on Tr 292M to BCN.	
BCN 1H 132.05	23	At LON 2d right onto Tr 278M to 'WOD'. At 'WOD' right on Tr 285M to ALY then Tr 282M to HTN then right on Tr 292M to BCN.	
BCN 1J 119.5*	10R	At LON 135R/2d(28L MM) right onto Tr 286M to 'WOD'. At 'WOD' left on Tr 285M to ALY then Tr 282M to HTN then right on Tr 292M to BCN.	
BCN 1K 119.5*	10L	At LON 119R/1.3d or 600(QFE), whichever is later, right onto Tr 286M to 'WOD'. At 'WOD' left on Tr 285M to ALY then Tr 282M to HTN then right on Tr 292M to BCN.	
BCN 1L 119.5*	05	At 23 MM (LON 097R/2d) right onto Tr 286M to 'WOD'. At 'WOD' left on Tr 285M to ALY then Tr 282M to HTN then right on Tr 292M to BCN.	

Revision: Nil

Fig. 11.5 SID route chart. (Courtesy AERAD).

from the ATIS and crews are normally aware of the standard instrument departure (SID) required for their route from a particular runway. Where a last-minute runway change is announced — e.g. changing of parallel runways for noise distribution, change of wind direction etc. — or an unexpected instrument departure is allocated, take-off calculations, radio selection, and rethinking of the departure may all have to be accomplished rapidly.

Today the crew can expect a 'Brecon one golf' SID from 28L, which is outlined on the SID route chart (Fig. 11.5). These routeings are normally memorized by crews but charts are also folded and clipped to the control column for easy reference. The departure routeing is to climb straight ahead on the runway heading to intercept the 261°R from London VOR and, at 7 DME distance from London along the 261°R, to turn right to track 274°M to the Woodley NDB then via ALY and HTN to Brecon VOR. The aircraft must cross Woodley above 4000 feet and level off at 6000 feet for further climb instructions. The departure frequency is noted as 132.05 MHz.

Arrival aircraft are restricted in descent to a minimum of flight level 70 at the holding points, thus maintaining separation. These SIDs not only feed the aircraft onto *en route* airways but are also designed to maintain aircraft clear of noise sensitive areas. The Captain selects the London VOR frequency 113.6 MHz on each receiver, checks the morse code identification, L-O-N, and sets 261 in each course indicator window. Woodley NDB frequency 357 kHz is selected on both ADF sets (ident W-O-D). The first height restriction of 6000 feet is set in the altitude window and the heading indicator on the compass is positioned to the precise runway direction of 276°M. Runways are designated by rounding the exact runway magnetic direction to the nearest ten degrees and then omitting the last zero. Hence 276 becomes 280 becomes 28, and where parallel runways exist left or right is added, as in 28L.

The First Officer and Flight Engineer continue the process of preparing the aircraft for flight while the Captain is now free to deal with problems as they arise. Preparing an aircraft for departure is a logistical exercise of considerable proportions involving much equipment and many people. Fuel is pumped aboard, engine oil and hydraulic fluid replenished, potable water tanks filled, toilets serviced, cargo and baggage loaded, galleys stocked, cabin interiors cleaned, minor faults repaired and, in winter, ice and snow removed from aircraft surfaces. Not suprisingly some difficulties do occur from time to time, and the Captain is kept fully informed of progress. Meanwhile, as time permits, each flight crew member arranges the charts required for the journey and checks beacons, instrument

Toilet Servicing Vehicle. Colloquially known as the Honeycart.

selections and INS way-points. Seats, headsets, and pilot's rudder pedals also require adjusting, and all take time. (When pilots' seats are positioned forward by the instrument panel for take-off, a good field of view is obtained over the glare shield with little impression being given of the vast bulk behind.) On completion of refuelling the Flight Engineer checks the figures with the refueller and passes the fuel book forward to the Captain for final check and signature.

Cargo loading.

Galley supplies.

With individual tasks complete the three crew members now come together for the first time as a team to begin the 'engine start' check. The time is approximately ten minutes to departure, and the Co-pilot reads the check list (Fig. 11.6) to which each responds as items are called. The list includes a final check by all on the INS present position and the first three way-points, radios are confirmed set and checked, and the correct altimeter setting selected. At the item on take-off speeds on the check list the check is halted while waiting for the arrival of the load sheet with the final take-off weight figure. At this point, with a few minutes to spare, the Captain takes the opportunity to give his briefing.

'Normal graduated power take-off. Departure is from two eight left, straight ahead to pick up the two six one degree radial to seven DME, then right to track two seven four degrees to Woodley. Above four thousand at Woodley, first height is six thousand. Transition altitude is six thousand. The wind is from the left so I'll have one notch down left on the ailerons. We may need engine anti-icing after take-off so I'll leave that to the Engineer to switch on if required. If an emergency occurs before V one I'll call 'abandon take-off', close the throttles and apply the brakes, manually select speed brakes and select idle reverse. If the emergency is an engine failure call me the symmetrical engines available*, and if the speed is above one hundred knots I'll select reverse power. If the emergency occurs after V one, take no action until the gear has been selected up, restate the emergency and I'll call for the appropriate drill. I'll fly the aircraft and handle the radio if you two carry out the drills.'

*Asymmetric reverse thrust is not approved.

ENGINE START CHECKLIST

Pre-Flight Check	Completed	ALL
Circuit Breakers	Set	E
Flight Deck Door	As Required	C
INS	Nav Mode	ALL
Oxygen Mask and Press	Checked	ALL
Window Heat	On	P
Emergency Lights	Guard closed, Light Out	C
Flt. Cont. Switches	Guards closed, Lights Out	C
Radios	Set	ALL
Pressure Altimeters	QNH set and cross-checked	ALL
Fuel on Board	....Kg	C/E
Probe Heat	Pitots only	P
Take-off Data	Checked indexed & bugged	ALL
Thrust & Start Levers	Closed and cut-off	C
Start Clearance (ATC)	Obtained	P
Seats & Safety Harness	Locked & Secure	ALL
Doors	Status....	E
Beacon	On	P
Galley Power	Off	E
Boost Pumps	On	E
Pack Valves	All Off	E
Clear to Start	Obtained	P/G
Brake Pressure	Pump ON and Checked	E
No.1 ADP or No.4 ADP	Auto (Pushback only)	E
Parking Brake & Press	Set and Checked	ALL
Start Pressure	Checked	E
Start Engines	4,1,2 and 3	E
Start Levers	Idle	C
Stabiliser Trim	Checked & Set	C/P
Nacelle Anti-Ice	As Required	C/E
APU Bleed Air Switch	Closed	E
Electrical Power	Lights Out, ess normal	E
Standby Power	Normal	E
Galley Power	On	E
Air Cond	Set	E
Bleed Valves	Open	C/E
Hydraulics	Auto and Normal	E
Aft Cargo Heat	Normal	E
Doors	Lights Out	E
Anti-Skid Ground Mode	Tested	E
Chocks & Grd. Equip	Removed	C
Dep. Time & U/C pins	GMT, U/C Pins Removed	C
AIDS	Insert	E
APU	Stopped	E
Brake Pressure	Checked	ALL
Engine Start Check	Completed	P

C - Captain   P - Co-pilot   E - Flight Engineer   G - Ground Engineer

Fig. 11.6 Engine start check list.

It is now six minutes to departure and a loading clerk presents the load sheet to the Captain as he finishes his briefing. The Co-pilot notes the actual take-off weight (304.9 tons) and stabilizer trim setting (6.0) and completes the take-off proforma (Fig. 11.7) while the Captain checks the loading. Speeds are shown as follows: V1 — 140 knots, VR — 158 knots, V2 — 164 knots, V2 + 10 — 174 knots, V2 + 40 — 204 knots, V2 + 80 — 244 knots. The difference between actual and maximum permitted weights allows a power reduction of 0.05 EPR, giving an EPR setting per engine for take-off of 1.39 EPR (maximum 1.44). With the load sheet checked and signed the clerk departs and the last door is heard to 'clunk' shut. The Captain and Co-pilot set the bugs on their airspeed indicators (ASIs) against the speeds noted from the take-off proforma, and EPR pointers are set on the engine gauges against the required EPR. With the take-off data set the thrust and start levers are checked closed and cut off and 'clearance' is called for start up. The Co-pilot selects 121.7 MHz on VHF box one.

F/O R/T - Clearance, good afternoon, Skyship One, stand kilo fourteen, with information 'sierra', request start-up.

No reply.

F/O R/T Clearance, Skyship One, how do you read?

(R/T has a readability code ranging from 1 to 5, 1 being virtually unreadable and 5 perfectly clear).

Clearance R/T: - Good afternoon Skyship One, reading you strength five, my apologies. Cleared start for Kennedy, Brecon one golf departure, squawk five three four two. Call one two one decimal nine for push back.

The First Officer acknowledges the start and reads back the clearance, which is as anticipated. Seats and safety harnesses are checked locked and secure, doors are checked closed, the anti-collision beacon is switched on and the Co-pilot requests clearance to start on intercom from the Ground Engineer (G/E). Parking brake is also checked and set, and start pressure checked as sufficient. Engines are started in order 4, 1, 2, 3 as No. 4 engine powers No. 4 hydraulic system, which supplies main brakes. Jet engine start is a precise sequence and the Flight Engineer liaises closely with the Ground Engineer on intercom during the procedure. The Flight Engineer announces 'Starting number four' and selects the ignition switch to 'ground start' to turn the engine. The Ground Engineer confirms on the intercom that N1 (the fan) is beginning to turn while the Flight Engineer monitors N2 rotation and engine oil pressure rising from the gauges. At 22 per cent the Flight Engineer calls 'Twenty-two percent' and the Captain places No. 4 start lever to idle while the First Officer starts the stop-watch. Twenty seconds is the

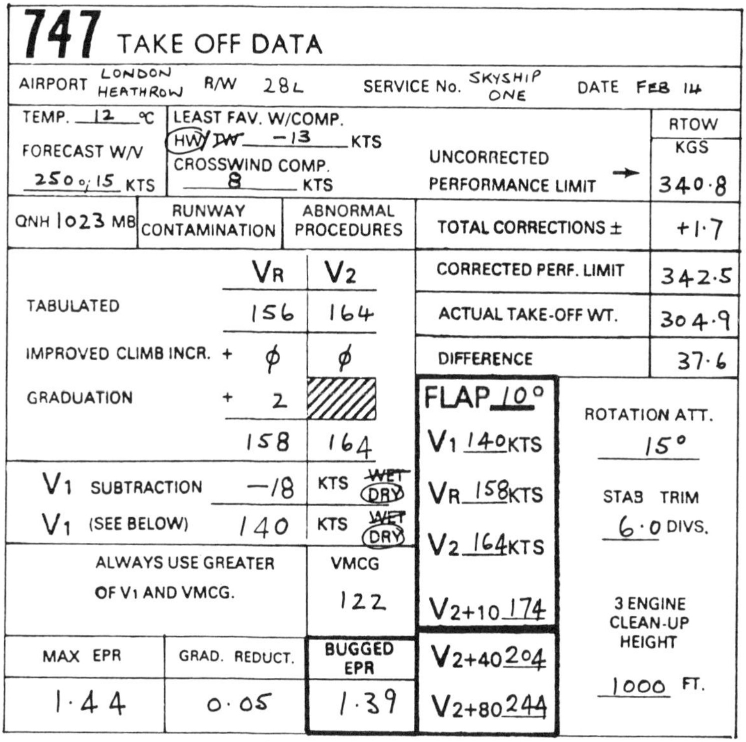

Fig. 11.7 Take-off Data.

maximum time allowed for start up. Fuel is now pumping into the engine and the igniters firing. Fuel flow is checked as normal. Five to ten seconds later the exhaust gas temperature (EGT) is seen to rise steadily, indicating light up.

On occasions the light-up process malfunctions and the EGT can be seen to climb rapidly towards the maximum of 650°C. A 'hot start' is called, and the start lever has to be placed quickly to the cut-off position to shut off the fuel and the engine turned for thirty seconds for cooling before a second start can be attempted.

This time, though, all is going well. The Flight Engineer calls 'Thirty per cent N two', and a check by all on No. 4 engine gauges indicates the engine winding up normally. At 50 per cent N2 the engine is self-sustaining and the Flight Engineer releases the ignition switch. The EGT continues to rise, peaks, then settles back to idling level. Engine instruments are checked for normal operation and warning lights are checked extinguished. No. 4 engine is now running. The Captain requests the First Officer to call for push back.

Push Back Truck.

At all stands with movable covered walkways aircraft have to be parked nose inwards to the building and are required to be pushed back before taxi. Push back trucks are wide squat vehicles loaded with heavy weights and push aircraft back via tow bars connected to the nose wheel. If insufficient space is available in front of the aircraft they are also designed to fit below the belly and pull aircraft back from underneath. Push back trucks are also used for towing around airport areas.

The First Officer switches to 121.9 MHz selected on VHF box one.

F/O R/T - Ground, good afternoon, Skyship One, kilo fourteen, request push back.
Ground R/T: - Skyship One, clear to push to face South.
F/O R/T - Skyship One, clear push, face South.

The Captain confirms with the Ground Engineer on intercom that all ground equipment has been removed and is clear of the aircraft, that chocks have been removed from the nosewheel and that the movable walkway has retracted.

G/E: - Release brakes, please.
Capt: - Brakes are released.

An engine roar is heard from the push back truck below the nose as the aircraft slowly begins to move backwards. The time is 1402 G.M.T. (Departures within three minutes of schedule are considered to be on time.) The Ground Engineer remains on intercom and walks by the nose wheel with the aircraft. No. 1 engine is started on push back. With the aircraft positioned for taxi, the Ground Engineer calls for brakes to be set to 'park' and engines 2 and 3 are started.

Starting procedures take a few minutes to complete, and when all engines are running the engine start check is continued. Start levers are checked at idle, stabilizer trim is checked and set, electrical, standby, and galley power are all checked normal, air conditioning is checked set, bleed valves are checked open, hydraulics are checked normal, door lights are checked out, the APU is checked stopped and brake pressure is checked normal. The Captain has a final word on the intercom with the Ground Engineer confirming the departure time of 1402 and requests the all clear signal on the left. The engine start check is now complete. The push back truck is detached and driven clear while the Ground Engineer disconnects his headset and stands safely to one side, arm raised vertically indicating all clear.

Capt: - Ask for taxi, please.
F/O R/T: - Skyship One, taxi.
Ground R/T: - Skyship One, clear taxi. Give way to the Pan Am seven four seven on your right, follow the outer taxiway for two eight left.

Around the central area are inner and outer taxying circles. As the Pan Am 747 passes a quick check on either side of the aircraft confirms that all is clear. The Captain releases the brakes and advances the thrust levers with the right hand, left hand on the tiller for steering. Power is kept to a minimum in manoeuvring areas to prevent damage to equipment and injury to personnel, the maximum allowable EPR being 1.05. Engine whine is heard to rise as the engines spool up and the aircraft moves off under its own power. Although graceful in the air the Boeing 747 is cumbersome on the ground; taxying the aircraft has been likened to driving a London bus down a narrow path while steering from the top deck. The maximum taxying speed in turns is 10 knots, and 5 knots in slippery conditions. The time is 1409 and Skyship One is on its way.

Taxi

The before take-off check is called for by the Captain as the aircraft taxies along the outer taxiway. (All check lists from now on are read by the Flight Engineer.) Take-off flaps are selected and checked set at 10°, speed brakes are checked down, flight controls are checked for operation and freedom of movement, flight instruments are checked, trim is checked set for take-off (stabilizer trim 6.0 divisons, rudder and aileron trim zero), annunciator panels are checked (all warning lights out) and pressurization is checked and set. The Captain also adds to his take-off briefing where necessary, and, if time permits, takes the opportunity to speak to the passengers to welcome them on board. In the cabin, departure preparations and emergency briefings are complete and the Chief Flight Attendant reports to the Captain to confirm the cabin ready for take-off, and that doors are set to 'automatic'. (With automatic mode selected the escape slides automatically deploy if the doors are opened in the event of an emergency evacuation.).

> *Ground R/T:* - Skyship One proceed down runway zero five then follow Pan Am to two eight left. Call tower now one one eight decimal five.
> *F/O R/T:* - Roger, Skyship One, down zero five, follow Pan Am and call tower. Good day.
> (The Co-pilot selects 118.5 MHz.)
> *F/O R/T:* - Tower, good afternoon, Skyship One is with you.
> *Tower R/T:* - Skyship One, roger. I'll call you back, you're number three.

Skyship One approaches the holding point behind Pan Am (callsign Clipper One), as a Saudi Tristar edges onto the runway and takes-off. Clipper One is cleared into position to hold. Pan Am is also going to New York and may affect Skyship One's chances of obtaining the requested track.

> *Tower R/T:* - Skyship One, after the departing Pan Am line up and hold two eight left.
> *F/O R/T:* - Skyship One, after Pan Am line up and hold.

Two minutes later Clipper One is cleared for take-off and, as the Pan Am 747 rumbles down the runway, Skyship One moves into position at the threshold. The Captain calls for the completion of the before take-off check, and the Flight Engineer makes the final PA call announcing take-off. Boost pumps are checked on, the fuel system is checked set, hydraulics and brakes are checked, air-conditioning valves are checked off, ignition swtiches are positioned to 'flight start' in case of engine flame-out on take-off and, with the aircraft positioned straight on the runway, the body gear steering is switched off. A final check from memory (using the mnemonic G-I-F-T-S) is made by each to confirm all is well. G—guards closed on instrument switches, I—instruments checked and set, F—flags retracted on all instruments and take-off flap set, T—trim set for take-off, S—steering (body gear) checked off. The aircraft is now ready for take-off as Skyship One holds on the runway for clearance.

Take-off and climb

In spite of the size of the large jets, advantage is still taken of prevailing winds and, where possible, aircraft land and take-off into wind. At heavy take-off weights even slight changes in the wind direction or speed can be critical. In hot climates temperature changes can also be a problem and even a 1°C rise can result in a two-ton reduction from maximum take-off weight. In today's conditions the wind of 250°M at 15 knots is blowing from 30° left of the runway resulting in a head wind component of 13 knots and a cross wind of 8 knots. Rotation speed is 158 knots relative to the air, so the aircraft will become airborne at a ground speed of 145 knots (158 minus 13), which can make a significant difference at heavy take-off weights, especially where the airport is hot and high. At rotation the ASI will, of course, read 158 knots.

The Boeing 747 can take-off and land with a tail wind (depending on the runway length and aircraft weight) but the maximum acceptable tail wind strength is only 10 knots. Cross wind components also affect the aircraft, the maximum acceptable cross wind on take-off being 30 knots gusting to 40 knots, and on landing, 25 knots gusting to 35 knots. Of course, these are maximum limits, and a 747 requires quite a lot of handling in such conditions. All motorists are aware of cross wind effects in open areas such as on elevated roads and suspension bridges, and high-sided vehicles can, on occasions, be overturned in strong winds. The effects are compounded with something the size of a 747 where the fuselage side acts like a giant sail, and even a cross wind of 8 knots will be noticeable. In all cross wind conditions aircraft tend to pivot on the undercarriage and weather-cock into wind, and in today's situation a degree of right rudder will be required to maintain the aircraft straight. In strong cross winds full rudder is applied opposite to the wind direction at the start of the take-off run and is progressively eased off to hold the aircraft straight as it gathers speed down the runway. Orange windsocks placed near the ends of runways are still a useful aid and give a clear indication of cross wind direction. A windsock being blown close to the horizontal indicates a wind speed of 25-30 knots. A further problem of cross-winds is that the into-wind wing tends to gain more lift than the wing in the lee of the aircraft fuselage and a degree of aileron is required to maintain the wings level. Today the Captain has already mentioned in his briefing that he estimates one division of aileron down left is required to be held by the Co-pilot at the start of the take-off roll.

The Pan Am 747 is now a few minutes ahead on his climb and Skyship One is given clearance for take-off.

Tower R/T: - Skyship One, cleared for take-off. Wind two five zero at fifteen.
F/O R/T: - Skyship One, cleared for take-off.

The Captain's right hand is on the throttles, the left hand on the steering tiller, and rudder is applied with the right foot. The aircraft is held stationary on the toe brakes. The First Officer holds the control column left aileron down and the Flight Engineer positions himself between the two pilots, slightly aft, ready to monitor power. The Captain calls 'Standby for take-off'. On brake release the Captain stands the thrust levers up to the vertical position, and elapsed time is started on the clocks. With throttle movement a take-off configuration warning horn sounds if a major item such as flaps or speed brakes are incorrectly set. Engine gauges are seen to wind up, and as the engines stabilize the Captain advances the throttles close to the required power setting.

Capt: - Set graduated power.

The Flight Engineer leans forward and finely adjusts each thrust lever to the precise EPR setting. The acceleration rate is rapid, with the aircraft gaining speed at 3-5 knots per second. As the ASIs become effective the needles are seen to move on both the Captain's and First Officer's instruments and the Co-pilot calls 'Airspeed building'. At 80 knots the rudder becomes effective and at the call of 'Eighty knots' from the Co-pilot the Captain transfers his left hand to the control column and continues to steer with rudder.

Capt: - I have the control column.
F/O: - Your control.

The controls feel uncomfortably crossed to the Captain with right rudder and left aileron. A few birds are flying near the runway edge and a starling is seen to pass below the nose of the aircraft and climb quickly to safety. (Birds at airports have always been a problem and all sorts of repellants, noise devices, and even birds of prey have been used to discourage them. Microwave emissions are known to disturb birds by somehow affecting their feathers and such equipment is presently under test.) Speeding down the runway the aircraft is steered along the centre line by delicate movements of the rudder pedals. The Captain's right hand is poised on the throttles ready to close them in the event of an abandoned take-off while his left hand on the control column holds down the windward wing. The First Officer and Flight Engineer scan instruments and monitor progress. At 140 knots the Co-pilot calls 'V one'. The aircraft is now committed to take-off and the Captain moves his right hand from the throttles to the control column, the Flight Engineer assuming control of the engines. Acceleration is still rapid. As the speed reaches 158 knots the Co-pilot calls 'Rotate'. The Captain pulls steadily backwards on the control column at just the right rate — not too slowly or lift off will be delayed, nor too quickly or the tail may strike the ground — until 15° nose-up attitude is reached on the attitude director indicator (ADI) and the aircraft is held in this position. At 1418 G.M.T. Skyship One lifts off and is airborne.

The B747 take-off run is around 35-40 seconds at normal weights but at maximum take-off weight it can be in excess of 50 seconds, using about three-quarters of the runway length. The Captain is now flying on instruments irrespective of the weather. 'V two' is called almost immediately by the First Officer and with the positive rate of climb indication, gear is selected up on the Captain's command. (In the gear-up sequence bay doors open first, increasing drag, before gear bogies retract into the bays.) At 174 knots (V2 + 10) the speed is held by adjusting the rate of climb of the aircraft. Passing 300 feet the cross-wind begins to bite and the Captain turns the aircraft into wind and crabs through the air along the extended centre line of the runway with the ailerons and rudder central. There is some light low-level turbulence and the aircraft is buffeting in the wind. Approaching 1000 feet the beam bar indicating radial 261° from London VOR is seen to move from the right towards the centre of the instrument and the Co-pilot calls 'Beam bar active'. The Captain positively turns the aircraft left and tracks down the radial holding the beam bar central in the instrument.

Gear Retraction sequence.

Tower R/T: - Skyship One, call departure one three two decimal zero five.
F/O R/T: - Skyship One, good day.
(The radio switch is flicked.)
F/O R/T: - Good afternoon, departure, Skyship One passing one thousand three hundred.
Departure R/T: - Good afternoon, Skyship One, maintain six thousand on reaching.
F/O R/T: - Skyship One, maintain six thousand.

Departure control confirms the height specified in the SID, or, if clear, assigns a higher level. At 1500 feet the Co-pilot calls the height and the Captain requests climb power. The throttles are eased back by the Flight Engineer to the climb power setting indicated on the EPR limit indicator, and the Captain reduces climb by lowering the nose to about 10°-12° attitude to allow for power reduction. Flight directors are switched on at this point and the altitude select switch is armed. The flight director pitch and heading bars are set as required and are used to help indicate the flight path as the Captain maintains handling control of the aircraft. The nose is now lowered a little more to gain speed for flap retraction. As the speed edges past 204 knots and is increasing, the Captain, calls for 5° flap. The speed is checked by the First Officer and 5° selected, with the Flight Engineer monitoring operation.

F/O: - Flap five set.

The aircraft is above 2000 feet now, barely climbing, and is still accelerating. At 7 DME on the 261° radial the Co-pilot calls the distance and the Captain turns the aircraft right to track to Woodley NDB (some 9 n.m. away), using the automatic direction finding (ADF) needles on the radio magnetic indicator (RMI) pointing towards the beacon. Just above 224 knots, the Captain calls for flap 1°. After a further check of the speed the First Officer selects flap 1°, and as the trailing edge flaps retract from 5° to 1°, half the leading edge flaps also retract automatically. With flaps extended beyond 1° a maximum speed limit of 240 knots applies and the Captain is required to maintain speed within the narrow 16 knots band between the minimum and maximum speeds of 224 knots and 240 knots (the margin is even less at heavier weights) by climbing the aircraft until 1° flap is set and acceleration can be recommenced. Concentration is also required in tracking towards Woodley.

F/O: - Flap one set.

The aircraft is now passing 3800 feet, 2 n.m. from Woodley and once again accelerating.

Capt.: - Flap in please.

Speed is checked above 244 knots and the First Officer selects flap zero. The remainder of the leading edge flaps retract automatically.

F/O: - Flaps are in.

Skyship One now enters cloud and continues to buffet in the light turbulence.

Capt.: - After take-off check please. The temperature is a bit low so I'll have engine anti-icing on, and better leave the seat belt signs on, for the moment.

As a precaution engine anti-icing is switched on when the temperature drops below 10°C in visible moisture. The appropriate switches are operated by the Flight Engineer and hot air is tapped from engine compressors to feed slots in the engine intakes to prevent ice accretion. Hot air can also be supplied to wing leading edges to prevent icing. At night, cloud penetration is not always apparent, but the switching on of a landing light illuminates the cloud and the situation is immediately obvious.

The Flight Engineer now commences the after take-off check. Passing over Woodley at 4300 feet (minimum height 4000 feet) the ADF needles swing round and point backwards. The Captain turns the aircraft right again to track 285°M out of Woodley and switches two of the INS sets to indicate cross-track and drift as a further check. INS is selected on the flight director and at 250 knots speed mode is engaged. The two yellow bars of the flight director on the ADI now indicate track centre line and attitude required to maintain 250 knots (in most terminal areas 250 knots is the maximum speed allowed below 10,000 feet). Passing 5000 feet the First Officer calls 'One to go', which is confirmed by the others.

Capt.: - Give him a call and say we're approaching six.
F/O R/T: - Departures, Skyship One is out of five for six thousand.

Departure control has Skyship One on radar with height indication but with a number of aircraft to control further climb clearance can be delayed.

Departure R/T: - Skyship One, maintain six thousand.

Obviously arrivals are preventing further climb. Still in cloud, the aircraft is climbing rapidly toward 6000 feet, and the Captain eases back on the power. The pilots confirm from the weather radar that the track ahead is free of Cb activity while the Flight Engineer completes the after take-off check. Probe heat is checked on, and ignition is left on in case of engine flame-out in the present turbulence. The temperature is below 10°C and engine anti-icing is already swtiched on, the no smoking signs are switched off (smoking during take-off and landing is a fire hazard and is not permitted), seat belts signs are left on as requested by the Captain, outboard landing lights are left switched on until 10,000 feet (a see and be seen precaution), landing gear is checked up and the lever checked to off, all lights out, air conditioning and pressurisation are checked and set, fuel systems are checked and set, and flaps are checked in, all lights out. The after take-off check is complete. Approaching 6000 feet a green light indicates the height has been captured.

All: - Altitude Green.

The Captain eases forward on the control column to maintain height and pulls back on the throttles to maintain 250 knots. The speed mode switch has dropped to off and the horizontal yellow flight director bar now indicates the attitude required to maintain 6000 feet. Just as the aircraft settles level at 250 knots departure control clears Skyship One for further climb.

Departure R/T: - Skyship One turn right heading two nine five for vectors to Brecon, recleared flight level one two zero. Call London one three three decimal six with your heading.
F/O R/T: - Skyship One, right two nine five, recleared level one two zero, London one three three six with the heading. Good day.

Skyship One is being vectored clear of incoming traffic by radar for further climb.

Capt.: - Climb power, please.

The Flight Engineer re-selects climb power and the Captain turns onto heading 295°M, easing back on the control column to climb the aircraft at 250 knots. The aircraft is now out of the London transition altitude of 6000 feet and the standard pressure setting of 1013.2 mb is set and checked on both altimeters.

F/O R/T: - London Skyship One, good afternoon, we're out of six thousand for level one two zero, heading two nine five.
London R/T: - Skyship One, good afternoon, maintain heading two nine five, recleared flight level one eight zero.
F/O R/T: - Skyship One, maintain heading, recleared level one eight zero.

Brecon VOR (frequency 116.3 MHz) is selected on both VOR receivers and identified. The inbound radial to Brecon, 292°, is selected on the course indicators, with the DME indicating 70 n.m., and the crew now have a pictorial view on the flight instruments of their position relative to track whilst under radar. Passing 10,000 feet the First Officer calls 'Altimeter check' and all acknowledge 'Passing ten for one eight zero, standard set'. Landing lights are now switched off. The Captain lowers the nose to accelerate to the normal climb speed of 320 knots and adjusts aircraft trim as required.

Settled in the climb at 320 knots the Captain engages autopilot 'A' with the heading mode selected to maintain 295°M and the speed mode engaged to hold 320 knots. Altitude select is armed to automatically capture flight level 180 on reaching. At 12,000 feet the aircraft breaks cloud into a bright blue sky. Ignition and anti-icing are switched off, the Captain cancels the seat belt sign, and passengers and cabin crew are free to move about. On VHF box 2 the Flight Engineer relays the airborne time to Company who signal New York with an estimate of arrival.

Approaching flight level 180 London clears Skyship One to flight level 280 and to proceed direct to Brecon VOR in South Wales. Brecon co-ordinates are in way-point 3 and 0-3 is keyed in the INS (present position direct to Brecon) and INS engaged. Navigation green lights glow on the pilots' panel indicating capture. 'Nav. green' is called by the crew, and the INS now automatically navigates the aircraft with autopilot 'A' engaged direct to Brecon, the direction being confirmed by the VOR. Passing 20,000 feet the second 'altimeter check' is called, and the INS indicates 9 minutes to go to Brecon, ETA overhead 1440, with the time now 1431. The aircraft has been airborne for just 13 minutes.

During climb the crew sum leg times on each flight log to obtain the estimated times at positions along route to JFK. The Co-pilot notes the estimate for Cork at 1503, and 53N 15W, the first position on track 'F', at 1536.

Capt.: - You'd better call Shanwick now for the oceanic clearance. I'll stay with London if you call Shanwick on box two.

127.65 MHz is selected on VHF box 2 and the First Officer calls Shanwick Oceanic control.

F/O R/T: - Shanwick, good afternoon, Skyship One.
Shanwick R/T: - Skyship One, good afternoon, go ahead.
F/OR/T: - Skyship One, out of Heathrow at one four one eight, estimating Cork at one five zero three and fifty three north fifteen

west at one five three six. Requesting track foxtrot, flight level three five zero, mach decimal eight four.
Shanwick R/T: - Roger, Skyship One, standby.

The First Officer maintains a listening watch on 127.65 MHz as the flight details are fed to the Oceanic Control computer for analysis. Meanwhile on VHF 1 London calls and clears Skyship One to continue climb to the initial cleared flight level of 310. The Captain acknowledges and selects 31,000 on the altitude select, which is confirmed by the Flight Engineer.

Shanwick R/T: - Skyship One, Shanwick, your oceanic.

The First Officer signals the Flight Engineer to monitor the clearance — with such important clearances it's imperative that two listen together.

F/O R/T: - Skyship One, go ahead.
Shanwick R/T: - Shanwick clears Skyship One, track foxtrot to Kennedy, flight level three three zero mach decimal eight four from five three north, one five west.
F/O R/T: - Skyship One is cleared track foxtrot, flight level three three zero, mach decimal eight four from five three north, one five west.

Obviously level 350 on track 'F' is not available.

Capt.: - Pity about the height. What other tracks are suitable?

The fuel flight plan indicates the next best track as 'echo', only two minutes longer. If flight level 350 is available on 'echo' a useful fuel saving could result.

Capt.: - Ask him about level three five zero on track echo.
F/O R/T: - Shanwick, Skyship One.
Shanwick R/T: - Skyship One, Shanwick, go ahead.
F/O R/T: - Skyship One, we'd prefer level three five zero if possible. How about level three five zero on track echo?
Shanwick R/T: - Roger, Skyship One, standby.

Climbing through level 290, at 500 feet per minute, Brecon passes below and the VOR needles swing round as the aircraft automatically turns to track 292°M, the centre line of airway UG1 to Strumble on the Welsh coast. INS way-points indicate 3 to 4 with Cork on way-point 5. Strumble is selected and identified on both VORs. During climb, power requirement progressively increases and is monitored continuously by the Flight Engineer who adjusts the thrust levers accordingly.

Shanwick R/T: - Skyship One, Shanwick.
F/O R/T: - Skyship One, go ahead.
Shanwick R/T: - Shanwick clears Skyship One track echo, flight level three five zero, mach decimal eight four from five four north, one five west.
F/O R/T: - Roger Shanwick, Skyship One now cleared track echo, flight level three five zero,mach decimal eight four from five four north, one five west.
Shanwick R/T: - That's affirmative. Continue now with domestic frequency, good day.

The final 'altimeter check' is called as the aircraft passes level 300. 'One thousand to go'. Indicated mach number increases with height (in fact the speed of sound decreases as the temperature drops) and on reaching the climb mach number of 0.82 the speed lock is disengaged and the aircraft is climbed using the vertical speed control to maintain M 0.82 constant. Cruise details extracted from a table against flight level 310 and the top-of-climb weight of 294 tons (10 tons is normally used in the climb) are set on the instruments. Engine cruise EPR of 1.32 and buffet speed (minimum cruise speed) of 252 knots are bugged. The required speed is noted as M 0.835. (Before departure, all-up weight is set on an indicator which runs down as fuel is consumed, and continually indicates total aircraft weight.).

F/O: - We've got level three five zero on track echo now.
Capt.: - Okay I'll continue with the radio if you sort out the paper work.

The aircraft is, of course, prepared for track 'F' and much rearranging is required. At flight level 310 the height lock captures and 'Altitude green' is called by all.

Capt.: - London, Skyship One now level three one zero.
London R/T: - Roger, Skyship One, level three one zero, call Shannon one three five decimal six.

Cruise

The time is now 1446 with the aircraft level over Strumble. The climb to level 310 has taken 32 minutes and covered 175 n.m. (At maximum take-off weight, over 200 n.m. is covered before cruise level 290 is reached from sea-level airports.)

Capt. R/T: - Shannon, Skyship One, good afternoon, level three one zero.
Shannon R/T: - Skyship One, good afternoon. We have you over

Strumble level three one zero. Do you have your oceanic?
Capt. R/T: - Skyship One, affirmative. Track echo, flight level three five zero.
Shannon R/T: - OK, Skyship One, continue to Cork at the moment. When you can give me an estimate for five four north, one five west let me know and I'll try to clear you direct.
Capt. R/T: - Roger, Skyship One, proceeding to Cork. Standby for the estimate.

Track 'E' should in fact commence at Shannon, but with clearances being issued by Shanwick to aircraft in turn as they arrive, Shannon control has the difficult task of rearranging the many aircraft onto their assigned tracks as they converge on the Atlantic entry points. Over Strumble Skyship One turns left to track 276°M along the centre line of airway UB10 to Cork. Cork VOR is now selected and identified on both receivers. At Mach 0.835 the Flight Engineer indicates cruise speed is achieved and the Captain calls for cruise power. Gently the Flight Engineer eases the throttles back to the cruise power setting of 1.32 EPR and the aircraft settles in the cruise.

The INS way-points from 6 onwards contain the position for track foxtrot, and the Captain keys the new positions of track echo into the INS. Track 'E' is noted as 54N 15W, 55N 20W, 56N 30W, 55N 40W, 52N 50W to Dotty, an imaginary position off the east coast of Newfoundland (Fig. 11.8). Way-point 6 is now position 54N 15W, and interrogating the INS with 0-6 reveals a time of 50 minutes from present position to 54N 15W. The time is now 1451 and the Captain passes the estimate of 1541 for 54N 15W to Shannon.

Meanwhile, the Co-pilot is rearranging the paper work. The fuel flight plan is not greatly affected as the new track lies only 1° north and is only two minutes longer, but the flight logs have to be

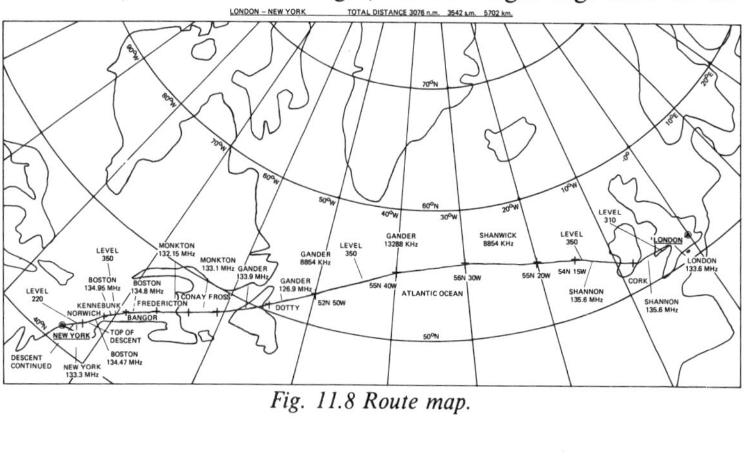

Fig. 11.8 Route map.

completely rewritten for the North Atlantic, and blank logs and logs for the expected re-routeing over North America are extracted from spares. Frequency 133.8 MHz is first selected on VHF box 2 for the track broadcast, and the First Officer checks the co-ordinates of track 'E' from the broadcast against those shown on the paper work on board. Having confirmed track 'E' the blank logs are completed by filling in the track echo co-ordinates and by noting the tracks and distances between them extracted from a tracks and distances manual. Ground speeds are noted from the original fuel flight plan and estimated times for each leg are inserted in the log. The completed paper work is then checked by the Flight Engineer and distributed among the crew.

Approaching Cork at 1459 Skyship One is cleared by Shannon control direct to 54N 15 W at flight level 350 and a further estimate is requested. The Flight Engineer informs the Captain that the aircraft is a bit heavy for level 350 and a request is made to maintain level 310 for the moment.

> *F/O R/T:* - Roger, Skyship One, cleared direct fifty four north fifteen west and estimating at one five three nine. We're a bit heavy and we'd like to remain at level three one zero for the moment.
> *Shannon R/T:* - Okay, Skyship One, maintain level three one zero. Cross five four north one five west level at three five zero and call me leaving and reaching.
> *F/O R/T:* - Skyship One, wilco, maintaining level three one zero.

The Captain confers with the Flight Engineer over the weight and a decision is made to climb to level 350 ten minutes before 54W 15W, while the First Officer checks the INS co-ordinates of track 'E' inserted by the Captain. On VHF box 2 the emergency frequency of 121.5 MHz is selected and a listening watch maintained while 123.45 MHz, the pilots' general 'chit-chat' frequency, is selected on VHF box 3.

At 1529 the aircraft weight is 286 tons and the Captain calls for climb power. The Flight Engineer advances the throttles and 35,000 is set in the altitude select window for automatic height capture. The aircraft is climbed at Mach 0.82 using the vertical speed knob, with the vertical speed indicator (VSI) showing a climb rate of 400 feet per minute. The new cruise power setting of 1.44 EPR and buffet speed of 255 are bugged, and the required cruise speed of Mach 0.84 on track echo noted.

> *F/O R/T:* - Shannon, Skyship One is out of level three one zero for level three five zero.
> *Shannon R/T:* - Roger, Skyship One, call level three five zero.

As the flight level 350 is captured 'altitude green' is called, and on reaching cruise speed of M 0.84 the thrust levers are retarded to the cruise power setting of 1.44 EPR. Skyship One is approaching 54N 15W now and the next way-point co-ordinates and INS track are checked again. A selection is made to display INS information on the instruments and the compass now indicates true heading and the beam bar the INS track. Over 54N 15W the aircraft automatically turns a few degrees left to track initially 291°T to 55N 20W.

> *F/O R/T:* - Shannon, Skyship One, five four north one five west at one five three nine, now level three five zero, estimate five five north two zero west at one six zero five, five six north three zero west is next.

Shannon Control repeats the times and estimates and instructs Skyship One to call Shanwick control at 20W on HF primary frequency 8854 kHz or secondary 13288 kHz. The First Officer selects and tunes HF2 on frequency 8854 kHz and calls Shanwick for a Selcal check. The equipment 'ding-dongs' a reply confirming operation. Headsets are now removed in the knowledge that the crew can be 'chimed' by Shanwick if required, and speakers are turned up to maintain a listening watch on 121.5 MHz. Hand mikes are available for communication and are convenient in the cruise with replies being heard on the speakers. Their use in flight is not permitted below fifteen thousand feet. Meanwhile the Captain is checking New York arrival times and lifts the PA handset to make his announcement to the passengers.

> *Capt: PA* - Good afternoon, ladies and gentlemen, this is the Captain. I trust you're comfortable and enjoying the flight. We're now level at thirty-five thousand feet, two hundred and fifty miles west of Ireland, with about five and three quarter hours left to go. We're estimating arrival in New York on schedule at four thirty-five in the afternoon, that's New York time, and for those of you who would like to set your watches the time in New York at the moment is ten minutes to eleven in the morning. The Atlantic route today takes us to fifty six north at thirty west, the mid-point of the crossing, making landfall over Newfoundland on the east coast of Canada. We then proceed across the Gulf of Saint Lawrence, down across New Brunswick, over the border into the States to overhead Bangor in Maine, down past Boston in Massachusetts, and on into New York. The *en route* weather should be good, although it may be a little bumpy from time to time at the other side, and the forecast for New York is overcast with fresh winds and the possibility of rain showers. If there's

anything we can do please don't hesitate to ask, and in the meantime I'll wish you all a pleasant trip.

Approaching 20W the INS way-points are once again checked for the next leg and at 1605 the aircraft automatically turns towards 30W. The time at 20W and estimate for 56N 30W are noted on the flight log and the Co-pilot transmits the position report on HF to Shanwick (after 30W Gander assumes control.).

F/O R/T: - Shanwick, Skyship One, eight eight five four, position. Five five north two zero west at one six zero five, level three five zero, five six north three zero west at one six five five, five five north four zero west next, temperature minus four seven degrees centigrade, wind two four zero diagonal three five.

In the meantime the Flight Engineer is completing his first fuel check while the pilots monitor the flight. The fuel used is summed from integrators and compared with the fuel still in tanks, as indicated by the gauges in kilogrammes, and the indicated all-up weight is also checked by adding fuel on board to the known zero fuel weight. Fuel to destination is calculated and subtracted from the fuel in tanks to obtain the fuel available on arrival. Obviously a minimum fuel figure at destination is required in case of diversion because of bad weather, and in adverse conditions of strong head winds or low flight altitudes, reserves may be reduced to a level at which an *en route* landing becomes necessary. Fuel condition is therefore monitored carefully throughout the flight and fuel checks are conducted at regular intervals.

The crew have been on duty for over three hours now, and with about 45 minutes to go to the next reporting point have the first opportunity to take a break. Two flight crew members must be seated and strapped in at all times to monitor flight progress and to cope with any emergencies that may arise, leaving one free to stretch his legs for ten minutes or so. At this stage, with the aircraft settled in the crossing, much of the flight is automatic, and crew duties consist mainly of monitoring. Navigation is checked at each way-point, flight and engine instruments are frequently scanned, systems are examined for normal operation and aircraft speed is constantly monitored. During cruise it is not necessary to keep an eye continuously on all instruments, as many, such as the engine instruments, are arranged in banks with pointers lying in line with each other. Any one out of line, or a small red failure flag on a flight instrument, sticks out like a sore thumb to crews who are familiar with every minute detail on the flight deck. Fuel, engine, and flight logs are also kept up to date, position reports radioed as required, destination weather copied at intervals from radio broadcasts and a wary eye kept for adverse weather along route. Obviously, flight crews cannot relax in the cruise to the extent of enjoying a good book, but while monitoring progress a magazine can be glanced at occasionally or the odd crossword attempted without adversely affecting the safety of the flight. Of course, conditions do change from flight to flight, and on certain journeys crews can experience little activity for some time, while on others they can be kept busy for most of the flight with weather avoidance, handling unexpected re-routeings, intense radio communications, the monitoring of bad destination weather reports, or re-planning flight fuel reserves etc.

In the tropics from Africa to Australia large Cb cloud can stretch up to 60,000 ft., and in active areas weather avoidance over long periods can be tedious and tiresome with the aircraft bumping about along the way. In the dense traffic of the North Atlantic aircraft normally have no choice but to maintain track and flight level and weather any turbulence, although severe conditions are rare in the region. In the remote areas of the tropics deviating 50 or even 100 miles from track to avoid large build ups of weather is not uncommon. During the day large Cb can be spotted quite easily if not in cloud, and at night the flashing and banging of thunder activity can be spotted hundreds of miles away. Peering at weather radar sets, with heading mode engaged on the autopilot, pilots weave their aircraft gingerly around the cloud cells to avoid the worst of the turbulence. Large Cb cloud can give an aircraft quite a shaking, and for the sake of safety and passenger comfort every effort is made to avoid them. (Maximum flight manoeuvring loads of the Boeing 747 are + 2.5 g to - 1.0 g, where 'g' is the force of gravity and negative 'g' the impression of weightlessness.)

On the North Atlantic routes, tracks do not follow great circle routeings but follow best time paths for prevailing wind conditions. While progressing from one position to another along track, however (e.g. from 55N 20W to 56N 30W), the INS routes the aircraft along a great circle (the shortest distance between two points on the globe) and the compass continuously changes heading as the aircraft maintains the great circle path. Over the continents, airways do not trace direct routeings but follow traditional routes from radio beacon to radio beacon placed at airports, towns, or along international borders. Many airways are short, and aircraft are required to twist and turn from beacon to beacon along route. Of course with INS engaged the aircraft flies great circle paths between beacons, but with many short legs little saving is achieved. Airways are up to 10 n.m. wide, and with INS engaged centre line tracking is generally very accurate.

In fact, before the days of INS, aircraft were more scattered owing to beacon fluctuations on airways and to the inaccuracies of conventional navigation in remote areas. Today's precise navigation results in opposite direction aircraft passing much closer to each other, with each accurately flying its track, and maintaining separation between flights is vital. Much of the world is without radar and a good lookout is the responsibility of the pilots. A bright red flashing anti-collision light is switched on at all times and can be seen from every angle. Strobe lights are also fitted on the wingtips of many aircraft and emit an extremely bright white flashing light which can be seen for many miles. At night navigation lights are switched on as an aid to sighting. The system is adapted from that used at sea with the left or port light being red, the right or starboard light green, and the stern light white. 'There is no red port left' serves as a reminder that red is port is left.

Although navigation and anti-collision lights may seem an anachronism in modern times they are still necessary for monitoring the movement of other aircraft in the vicinity, especially at night where radar is not available. When the lights of another aircraft are seen to move in relation to an observer the two aircraft should pass safely, but when the observed lights appear stationary there is a definite risk of collision, and a change of course may be necessary. Landing lights are also a useful anti-collision aid and are not only used for taxi, take-off, and landing, but at all times below 10,000 feet, even during the day, they are switched on as a 'see and be seen' precaution. In the cruise, landing lights are often used to signal other aircraft. On airways at night, especially in remote areas, opposite direction traffic flash landing lights at each other as an added safety precaution. To add to the confusion some stars and planets distinctly twinkle red and green and can sometimes be mistaken for a distant aircraft. There's more than one pilot who has found himself flashing at a star! On occasions opposite direction traffic can be so dense that the flashing of a landing light can result in a dozen aircraft flashing back in the distance.

In certain areas direct routeings off airways are available, but only where radar coverage is good, such as in central Europe and North America. When on airways, beacons are selected in turn along route to check INS progress, and in sensitive areas from the Balkans to the Far East aircraft are required to adhere precisely to airways systems. Radio reporting can be tedious and formal, and aircraft are often required to obtain permission before entering airspace. On occasion two, and even three air traffic controllers are in contact simultaneously, for instance over Cyprus, where Turkish, Cypriot and Beirut controls all require position reports, and none is in touch with each other.

Certain journeys require re-routeings that are known in advance, such as flights from Europe to South Africa which must avoid Libyan airspace and route via Egypt, adding time and miles to the journey. In war-sensitive zones such as the Middle East or Persian Gulf unscheduled re-routeings or flight level changes are not uncommon, and all add to the workload of the crew. At times, for example, when the situation in the Lebanon gets very hot, Beirut control can disappear altogether! ATC strikes and controllers working to rule also add to the strain, and recent years have seen disruptions in the UK, France, Italy, Scandinavia and the United States. Delays inevitably result in lengthened duty days for crews.

Destination, diversion, and *en route* airport weather forecasts and actual reports are broadcast on HF at regular intervals (every 30 minutes) and, with severe weather expected for the arrival period, reports are copied throughout the flight (if broadcasts can be received) in the hope of improvement. With the threat of a diversion because of bad destination weather, flight logs are prepared and fuel requirements calculated for the diversion airport, or perhaps somewhere more suitable, depending on weather reports. When a busy destination airport closes because of bad weather the organized chaos that ensues is unbelievable, with everyone scattering for open airports at the same time, and it's as well to be prepared for any eventuality!

* * *

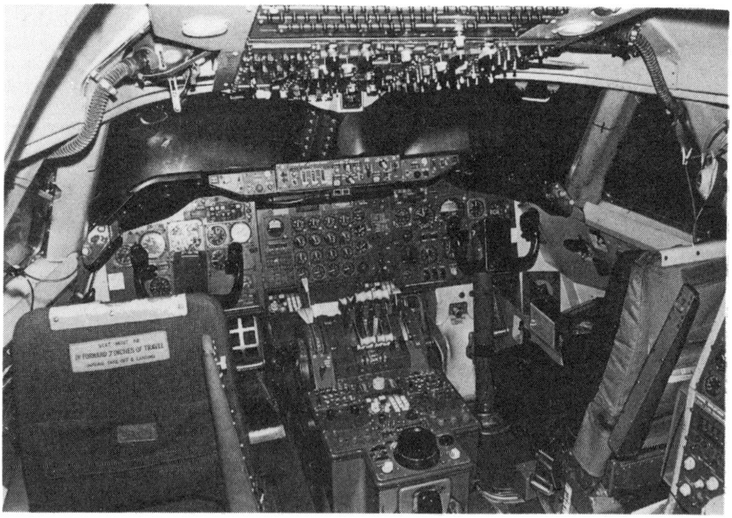

Boeing 747 Simulator Flight Deck.

Boeing 747 Control Cabin

Although considerably larger, the control cabin of a 747 is less complex than that of a 707. The major portion of the instrumentation in a 747 control cabin is identified in this picture of a Trans World Airlines superjet cockpit:

1 Overhead switch panel
2 Standby compass
3 Navigational radio selector
4 Autopilot engage switch
5 Navigation mode selector
6 Speed mode selector
7 Airplane registry no.
8 Map, panel light knobs
9 Central instrument warning lights
10 Approach progress indicators
11 Inertial navigation warning lights
12 Thrust reverser indicator lights
13 Pitch trim controls
14 Clock
15 Mach/airspeed indicator
16 Gyro horizon
17 Electric altimeter
18 Radio altimeter

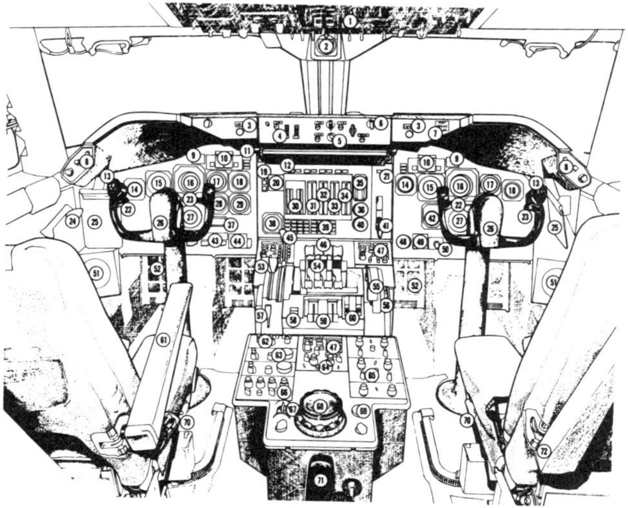

19 Navigation marker beacon lights
20 Total air temperature
21 Gear down/locked indicators
22 Microphone switch
23 Autopilot disengage
24 Nose gear tiller
25 Chart holder
26 Pilot's control yoke
27 Horizontal direction indicator
28 Vertical speed indicator
29 Standby altimeter
30 Engine pressure ratio gauges
31 Low speed engine comp.
32 Exhaust gas temp.
33 High speed engine compressor RPM
34 Fuel flow indicator
35 Flap position indicator
36 Static air temp.
37 Turn and bank indicator
38 Flight control position indicator
39 Annunciator light panel
40 True airspeed indicator
41 Landing gear control handle
42 Radio magnetic indicator
43 Instrument switches
44 Reserve brakes
45 Speed brake handle
46 Thrust levers
47 Inertial navigation controls
48 Brake pressure
49 Water injection control
50 Computer selector switch
51 Weather radarscope
52 Rudder pedal
53 Manual stabilizer trim levers
54 Go-around switches
55 Flap lever
56 Pitch trim
57 Parking brake latch
58 Parking brake light
59 Engine start levers
60 Stabilizer trim cutout switches
61 Seat arm, down position
62 Weather radar control panel
63 Automatic direction finder
64 Air traffic control transponder
65 Very high frequency radio
66 Ultrahigh frequency radio
67 Aileron trim
68 Rudder trim
69 Warning horn silencer
70 Seat positioning controls
71 Pilot's handset for intercom and passenger announcements
72 Seat arm, up position

Boeing 747 Control Cabin. (Courtesy Boeing Airplane Co.).

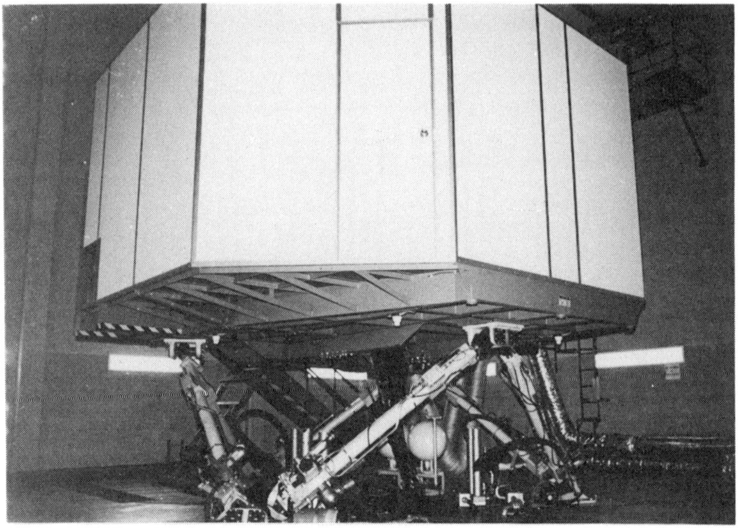

Boeing 747 Flight Simulator.

At 1656 G.M.T. 'oscar foxtrot' turns over 56N 30W, the mid-point of the North Atlantic crossing, and heads, initially, 264°T on the great circle track to 55N 40W. Cork now lies just under two hours behind with the coast of Gander two hours ahead. On each Atlantic crossing an equal-time point, which takes account of the wind, is calculated between suitable onward or return airports in case of emergency in mid ocean. In operations before departure the weather at both Shannon and Gander was reasonable, and the equal-time point between the two calculated as 1710 G.M.T. If an emergency, such as an engine failure, occurs before this time, it will be quicker to return to Shannon, and after it, to continue to Gander.

In the present day, however, emergencies are a rare occurence. Modern equipment is extremely reliable and malfunctions uncommon, although crews are trained to cope with any eventuality. To maintain standards flight crews are required to practise emergency procedures twice a year in a simulator, a machine on the ground which, quite simply, simulates flight. Early simulators used to be disparagingly referred to as 'cock ups of mock pits', but today's models are exact working copies of a particular flight deck and are highly sophisticated pieces of equipment. Simulators now have freedom of movement in all directions, with very realistic visual displays.

At one time all airline training was completed on aircraft, but with more accidents occurring from practice engine failures than from actual engine failures, training was progressively transferred to simulators. In airlines today, all pilot training (whether new entry or converting from one type to another), including instrument flight checks and flight deck procedure refreshers, are conducted on the simulator. Experienced fast jet pilots on conversion courses, say converting from the Boeing 727 to Boeing 747, require only a few hours on the aircraft after simulator training before going on service, and simulators are now so sophisticated that approval has been given in the United States for complete conversion training on advanced simulators, with zero actual flight time on type, before commencing route flying under supervision.

The simulator offers training that is effective, safe, and cheap. A lifetime's experience of flying can be achieved in a comparatively short time on the ground, with flight crews being trained to cope with situations that may never be encountered in flight because of the high reliability of modern equipment. However, in spite of the usefulness of simulators in such circumstances, it is still recognized that there is no substitute for actual flying experience. Personnel may sweat during a difficult exercise in the simulator, but it's a different kind of sweat. In the air it is the problems of the real flying world that make the adrenalin flow. There is no safer pilot than one who has, on occasions, been a little afraid.

On board all the big jets flight data recorders (FDRs) record a number of flight parameters throughout the flight, and cockpit voice recorders (CVRs) record flight deck conversations on a continuous half-hourly basis. The FDR is situated in the tail, considered the strongest part of the aircraft, and the recorder is built to withstand an impact of 5000 lbs, 100 g, and a temperature of 1100°C. (This is the so called 'black box', which in fact, is coloured orange to aid recovery). Details of incidents and emergencies are therefore carefully recorded, and can be analysed at a later date. Visual and aural warnings, which alert crews in the event of malfunction or failure, abound on the flight deck, and lights (flashing and steady), bells, horns, clackers, wailers, tones, and voice warnings illuminate or sound to warn the crew of danger. Warning lights grouped on annunciator panels illuminate amber for minor malfunctions and red for more serious faults. (Each warning light contains two bulbs in case of bulb failure.) All procedures, whether fault finding, precautionary, or emergency are completed meticulously from check lists, but emergency drills requiring immediate action are initiated from memory. However, the image of the fast reactioned jet pilot speeding into action and racing through drills is far from true. In most cases situations are studied carefully, and perhaps even discussed at length among the crew, before action is taken, one of the few exceptions being engine failure on take-off when quick thinking and fast reactions are essential.

Engine fire is potentially the most serious incident and with such an event a bell rings and a red light illuminates on the associated engine fire handle. On command from the Captain the First Officer and Flight Engineer initiate the fire drill from memory while the Captain maintains control of the aircraft and handles the radio. The associated thrust lever is closed, the start lever is placed to cut off, effectively shutting off the fuel, and the fire handle is pulled, disconnecting systems from the affected engine and arming the appropriate fire extinguisher system. Shutting off the fuel normally extinguishes the fire, but if the fire handle light remains illuminated indicating that the engine is still on fire, then a fire bottle containing extinguishant is discharged into the engine, and if required a second bottle is discharged thirty seconds later. It may also be necessary, on the rare occasion, to increase speed to blow out the fire.

An emergency such as engine failure on take-off requires split second decision making and positive action. Engine failure at the high power settings required for take-off usually results in the engine going with quite a 'bang'. The go or no go decision speed is annotated V1, and up to this speed sufficient runway is available for the aircraft to stop. After V1 there is insufficient runway remaining and the aircraft is committed to take-off. Abandoning take-off close to V1 is a hazardous procedure and the decision is not taken lightly. When a serious emergency arises, i.e. engine failure, before V1, the Captain immediately calls 'Abandon take-off', closes the throttles, deploys the speed brakes, selects reverse thrust and at the same time applies full brakes. Taking-off at maximum weight the V1 speed can be close to 145 knots (165 mph) and stopping from such a speed generates enormous brake heat energy. The brakes almost certainly overheat, with a resultant fire risk, and require hours to cool. The tyres are fitted with fusible plugs which melt at 350°F to prevent explosions.

One emergency in the air which may require swift action is instantaneous loss of pressurization owing to system failure, window blow out, or the fuselage skin being punctured. As the cabin depressurizes moisture in the cabin air condenses instantly because of the rapid drop in pressure, producing a fine mist that reduces visibility in the cabin. Above about 10,000 feet, human performance begins to deteriorate through lack of oxygen, although sufferers gain a false sense of confidence in their abilities. This condition of shortage of oxygen to the brain tissues is known medically as hypoxia. On instantaneous exposure to the rarefied atmosphere at 35,000 feet, the time of useful consciousness is one to one and a half minutes. When such a situation arises, masks in the cabin drop automatically from ceilings for passenger use, the Flight Crew don oxygen masks, and an emergency descent is initiated. The oxygen is sufficient to supply passengers and crew for some time at lower altitudes but is rapidly used up at height. Also, any failure of the oxygen system would, of course, prove disastrous, so descent action is initiated promptly.

The Captain closes the thrust levers, selects speed brakes, and waits for the speed to reduce below the maximum operating speed for gear lowering of 270 knots. The gear is then lowered and the autopilot disconnected. The Captain turns the aircraft off track and pushes the nose down sharply until maximum speed is achieved. In this condition, with so much drag, the aircraft descends at 10,000 - 12,000 feet per minute. At somewhere between 10,000 and 14,000 feet, perhaps higher depending on the safety height, the aircraft is levelled off and a landing made at the nearest suitable airport. Below 14,000 feet passengers are free to breathe without oxygen masks, but the flight crew must remain on oxygen supply until the aircraft descends below 10,000 feet.

Emergency procedures also include cabin fire drills (BCF extinguishers for all types of fire and liquid extinguishers for damping deep-seated fabric fires are positioned at numerous points throughout the aircraft), cargo hold fire drills, flight deck and cabin smoke removal drills, two engine approach and landing drills, partial main gear landing drill (i.e. one failed to lower), system malfunctions, flap, stabilizer and flight control failures, and many others. The check list containing normal, alternate, and emergency drills is a book in itself.

Flight crew incapacitation is a problem of which pilots are aware, and procedures are practised in the simulator. Such situations are rare, although incidents of medical complications or food poisoning have been known to take their toll. (All flight deck food is prepared by a separate caterer, and the Captain and First Officer eat different meals at different times.) Two flight crew members can handle the landing of a Boeing 747, but this is not done under normal conditions. An emergency is declared to air traffic control, and the flight crew are given all the help and time required. On one occasion a Captain suffering from diarrhoea in flight was advised by a doctor travelling as a passenger to take three pills to ease his frequent visits to the toilet. The Captain was reluctant to accept medicine but under pressure from the doctor finally agreed to swallow two. However, by mistake, the doctor had given the pilot sleeping tablets, and the Captain was soon snoring soundly in his seat, only to wake several hours later in the airport medical centre after landing!

With flight control malfunctions, major system failures or

undercarriage problems, a crash landing can be considered a possibility and an emergency declared. All services — fire, police, rescue, hospitals — are alerted. (Exercises are conducted at intervals by these organisations to monitor the capacity of coping with such events.) Crash landings as such are rare, and the term is more usally applied to landing in an emergency situation, where assistance is required when the aircraft is safely down. For example, a main gear indicating 'not locked down' requires initiation of the crash landing drills and alerting of the authorities, but more often than not results in the aircraft landing safely without trouble. A blazing engine fire on landing requires fire services and, once stopped, an emergency evacuation of the aircraft will almost certainly be necessary. In fact, fire after landing from engines, or ruptured fuel tanks can often be the greatest risk in an emergency and in such cases evacuation from the aircraft is usually required to be carried out as swiftly as possible. Tests have shown that if a fuel tank is ruptured on impact the ejected fuel is sheared by the airflow into a highly flammable mist. A polymer additive with a high molecular weight is now being developed to act as an anti-misting device, which, it is hoped, will greatly reduce the risk of immediate post-crash fuel fires in otherwise survivable accidents. Landing gear up, or partially lowered, is a distinct crash landing situation, and at certain airports foam blankets can be laid on the runway to smother sparks and reduce the risk of fire.

In an emergency situation, with the prospect of an emergency evacuation after landing, passengers are instructed to prepare for such eventualities. Spectacles, ties, false teeth, and shoes are removed. Seat backs are positioned upright to prevent passengers behind from striking the seat in front and to protect the occupant from backlash after any impact, and tables are stowed. During all take-offs and landings it is imperative that hand baggage is stowed below seats to ensure aisles and door openings are kept free. The crash landing drills call for the Co-pilot at 1000 feet from touch down to inform passengers on the public address system to take up landing positions. Passengers are instructed to lean forward with seat belts tightly fastened, placing their heads on cushions on their laps with arms folded across their heads for protection. At 200 feet the Co-pilot calls 'Brace, brace' and passengers tense for landing. If the emergency is an unsafe landing gear, or engine or cabin fire, the risk of fire spreading can be anticipated, and it may be prudent to initiate an emergency evacuation immediately the aircraft has stopped. The brakes are first set, the aircraft depressurized, and the engines shut down. The passenger evacuation signal is activated, and cabin crew open doors that automatically deploy escape slides. Passengers are instructed to jump and sit on the slide in a continuous flow, and a jumbo full of passengers can be evacuated in 90 seconds (assuming only half the exits available) by such a system.

Crash landings have also to be anticipated and planned for when aircraft are away from airports in the more remote parts of the world, and trans-Polar flights, for example, carry Arctic survival equipment including polar suits, kerosene stoves, and snow shovels, etc. Seats are removed at the rear of the cabin and Arctic survival euipment is lashed to the floor under a restraining net.

With much of the earth's surface covered by water the prospect of ditching is a possibility, although nowadays extremely rare, and life jackets and slide/rafts (Fig. 11.9) are carried on all overseas flights. A life jacket is normally placed below every seat and, at each main door except the overwing exits, an escape slide/life raft combination can be automatically inflated when the door opens. Once afloat in the open sea a canopy can be fitted over the uprights on either side for protection. At the overwing exits life rafts and survival packs are positioned in overhead lockers.

When landing on the sea, the flight crew are instructed to attempt to land the aircraft along the line of the primary (i.e. predominant) swell, and upwind and into any secondary swell, or downwind and down the secondary swell. The aircraft is landed with full flap at normal speed, but with the gear up to present a smooth fuselage to the water surface. After ditching, with the aircraft settled on the water, the doors are opened and the slide/rafts automatically deployed. At overwing exits life raft lines are attached to a door hinge and each life raft pack tossed into the sea uninflated. Once in the water a sharp pull on the line inflates the life raft and evacuation can begin. With passengers, crew, survival packs, emergency beacons (which emit a swept tone on the emergency frequency of 121.5 MHz when the antenna is erected) and first aid kits aboard, the slide/rafts and life rafts can be detached and allowed to drift from the sinking aircraft. Once free of the aircraft, attempts should be made to lash rafts together in a group to aid rescue.

Two further eventualities for which crews are trained are hijackings and bomb scares. The first case of unlawful interference with a civil airliner occured in Peru in 1930 when a group of revolutionaries commandeered an F-7 to flee the country. It was not until the late forties and early fifties, however, that hijackings were recorded with any frequency, when about one dozen successful acts of air piracy involving commercial airlines were committed by people fleeing from Iron Curtain countries. After a lull, the late fifties saw the beginning of the 'take me to Cuba' hijackings, which spread the crime of air piracy worldwide to epidemic proportions, culminating

Fig. 11.9 Slide/raft.

in 70 successful and 19 attempted hijackings in 1969, the largest number to date in any one year.

In 1970 began the hijackings of the Popular Front for the Liberation of Palestine (PFLP) to obtain release of Arab guerrillas in European jails. On 6 September 1970, three aircraft, a TWA Boeing 707, a Swissair DC8, and Pan Am Boeing 747 were hijacked simultaneously by the PFLP, and three days later, on 9 September, a BOAC VC10 was hijacked by the same organization. The Pan Am Boeing 747 was destroyed in Cairo the day after being hijacked, while the remaining aircraft (B707/DC8/VC10) were flown to Dawson's Field, an abandoned airstrip in Jordan, and held for a few days before being systematically blown up on 12 September 1970. The passengers were detained by the terrorists for a further three weeks while governments negotiated the release of prisoners and this resulted in the freeing of a total of seven Arab guerrillas by Switzerland, West Germany and the United Kingdom. Thus ended the blackest month in the history of aviation hijacking.

In the early seventies the hijackings for financial gain began, in which enormous sums were demanded as ransom for the safe release of passengers, crew and aircraft. Almost all were unsuccessful, except one in particular, perpetrated by a middle-aged man known as D. B. Cooper, who became something of an American folk hero. On 24 November 1971, the so called Mr Cooper threatened to blow up a Boeing 727 in mid-air unless he was given $200,000 and two parachutes. The aircraft landed at Seattle airport, just long enough for his demands to be met, and shortly after take-off Cooper jumped out, never to be seen again.

The late seventies and early eighties have seen a marked drop in hijackings, as not only is the crime out of vogue, but havens for hijackers have diminished considerably as Third World governments have become less tolerant of such behaviour. Security at airports has been tightened extensively, and metal detectors, the searching and X-raying of hand baggage, and the identifying of individual suitcases have all helped to improve safety, not only with regard to hijackings, but to bomb scares as well. The authorities, too, are much more able to cope with hijackers, and the risk to such criminals is now so high that imprisonment usually results. On the rare occasion culprits are slain by security forces.

Airlines and countries adopt various attitudes to combating terrorism in the air. In the United States, the Federal Aviation Authority (FAA) forbids passengers to visit the flight deck during flight, while in the USSR crews are encouraged to physically resist hijacking attempts. In some airlines the flight deck door is locked at all times, and armed sky marshals are carried on flights, although a shoot-out with terrorists in mid-air could prove disastrous if a bullet pierced the fuselage skin. Others have a more relaxed attitude to flight deck access, believing, not unreasonably, that determined terrorists armed with modern weapons could easily overpower an unarmed crew by force, and that the line of least resistance is the safest approach for passengers. In fact, it is extremely rare for hijackings, although sometimes bloody and violent, to result in an aircraft accident, and past experience has shown that once overpowered, safety can best be maintained by crews complying with terrorists' instructions. Whatever the approach in the air, strict security precautions at airports seems to hold the answer, and in the future, when all governments refuse to harbour terrorists, hijackers will have nowhere to go except prison.

Incidents of bomb warnings occur on the rare occasion when an airline is informed that an explosive device has been placed aboard a particular flight. Although taken seriously they usually turn out to be hoaxes. A bomb warning received when the aircraft is on the ground can be dealt with swiftly as any passengers on board can be quickly evacuated and a search conducted immediately. In the air it is a different matter, and the aircraft has to be landed at the nearest suitable airport. *En route* to the diversion airport the aircraft is searched discreetly and the passengers prepared for an emergency evacaution after landing. If a suspicious device is found, those sitting nearest can be moved and the suspected bomb covered with blankets, seat cushions, and hand baggage to help contain any explosion. As a further precaution the aircraft is also descended and depressurized to reduce the risk of the fuselage skin being punctured if detonation occurs.

* * *

Skyship One crosses 55N 40W at 1741 G.M.T. and proceeds towards the next track position of 52N 50W. The temperature is showing -51°C and the wind 245°T at 40 knots, giving a ground speed of 463 knots. The Captain disengages the autopilot to check aircraft trim, and on re-engagement notes the nose-up attitude of $2\frac{1}{2}$° on the attitude director indicator (ADI). An added advantage of this nose-up attitude is that the underside of the aircraft produces positive lift and additional fuel savings can be achieved. The navigation check confirms INS way-points and tracking, the flight log is updated with the 50W estimate, and the position report transmitted to Gander on 13288 kHz.

F/O R/T: - Gander, Skyship One on one three, position.
Gander R/T: - Skyship One, Gander, go ahead.
F/O R/T: - Skyship One, five five north four zero west at one seven four one, level three five zero, five two north five zero west at one eight three four. Dotty is next.

Gander repeats the message and instructs Skyship One to call at 50W on 8854 kHz.

F/O R/T: - Skyship One, that's all affirmative Gander, and to call at fifty west on eight eight five four.

At some time during the flight the crew require a meal which is generally served after the passengers have been attended to. Meals are eaten from trays on laps. Also, during cruise, some airlines permit visits to the flight deck at the Captain's discretion, and occasionally passengers are invited to view, although not, as mentioned, in the USA where FAA rules forbid such visits. General first impressions are that the flight deck is small and is noisier than expected with a background rush of air being heard. Most also comment on the apparent lack of movement with clouds far below seeming quite stationary.

The Boeing 747 flight deck instrumentation is not the most advanced, as the aircraft was designed back in the late sixties and first entered commercial service in 1970. Modernization programmes have, of course, been introduced and more sophisticated equipment retro-fitted on a continuous basis. The basic instrumentation, however, remains an analogue and digital presentation with many dials and pointers. Advanced flight decks have managed to lose a cluttered appearance and displays now consist mostly of digital readouts on TV-type cathode ray tube (CRT) screens. Flight deck design on the Boeing 757, 767, and the Airbus A310 incorporates full use of CRT displays. The attitude director indicator and the horizontal situation indicator are no longer mechanical and electrical instruments with moving parts, but are computer-generated pictures presented on small screens.

Inertial navigation system (INS) computer navigation now has the addition of flight management system (FMS) computers which provide information on aircraft performance, fuel management and navigation data. Details are presented in digital form on a small display screen, with the facility to select pages as required. Information is inserted and requested by pressing numbered keys similar to the INS. Information on flight details can be based on best time, lowest cost, maximum range or minimum fuel, but the obvious advantage of FMS is in its fuel-saving capacity. The FMS receives continuous inputs of altitude, ground speed, fuel flow, engine pressure ratio, N1 (fan) speed, and aircraft configuration.

In the performance mode, optimum height, speed, engine power setting in the cruise, and energy efficient profiles for all phases of flight can be displayed. In the navigation mode, navigation inputs from the computer are displayed. Before departure a specific route number is inserted and the aircraft is then automatically navigated along the required route. The FMS also automatically selects and tunes VOR/DME beacons *en route* and updates the navigation information accordingly.

Use of FMS in the cruise allows automatic operation of the thrust levers via an autothrottle computer to maintain desired power settings and speeds. Minor variations in speed are controlled through the autopilot by an energy method that exchanges airspeed for height. Small fluctuations in airspeed caused by gusts are converted into height by slightly climbing or descending the aircraft with increase or decrease in airspeed, and is achieved with only minimal variation in height (plus or minus 50 feet). The best flight profile is thus achieved without continual minor thrust lever movements which are wasteful of fuel. The autothrottle computer is also programmed to compute the aircraft altitude required in the event of an engine failure, and to protect the engine from stall and flame-out.

On take-off some autothrottle systems can be engaged to advance the thrust levers automatically to the preset take-off power setting. Thrust is automatically reduced in the climb on selection of climb mode, and the aircraft can be climbed at constant power setting, airspeed, or Mach number, depending on requirements. On the approach the autothrottle computer sets the power required for the flight profile selected by the pilots, and automatically closes the thrust levers at touch-down on automatic landing. If a missed approach results, go-around power is automatically set by the pressing of a go-around button on the thrust levers.

The advanced systems described above are in daily use on the latest jets, and the flight decks of the future, or at least for the next decade, are with us today. 'Head-up' displays, for example, are common on fighters, and are being developed for commercial use. With these the pilot can maintain his head in the normal look-out position for landing while information from the main flight instruments is displayed on the windshield in front of him; especially useful in adverse weather conditions. Glide slope and runway centre line guidance would also be displayed, perhaps in the form of an artificial runway, which would merge with the position of the actual runway as the aircraft breaks cloud.

On the ground, microwave landing systems are already installed at certain airports and will eventually replace the radio signals of the ILS. Microwave landing systems are not subject to interference and allow curved approaches to be flown to landing. Flight control systems, known as 'fly-by-wire', are now in use and operate by electrical signals without any mechanical connections between flying controls and control surfaces. When a pilot commands a flight condition change by movement of the control column the required control surface movement is determined by computer. Such systems can be programmed to reject pilot flying control demands that may compromise the safe operation of the flight. Changes to pilot flying controls are also envisaged, and already an Airbus A300 is flying on test with a mini flying control side-stick by the Captain's left knee in place of the standard control column. The mini stick is only 12 cm long and allows the aircraft to be flown by one hand. Removal of the standard control column also improves flight instrument view.

In navigation, laser gyroscope navigation systems are in daily use, and satellite navigation is being developed which should be available for aircraft navigational purposes by the 1990s. One system, Navstar, already has nine satellites transmitting in orbit, and even today light navigational equipment is available for yachtsmen. Twenty-four satellites in all will be required for full navigational coverage across the world, and when complete will navigate aircraft worldwide to an accuracy of twenty feet.

Aircraft size is not likely to increase over the next decade, but already the Boeing 747 upper deck is stretching further and further back. The present Boeing 747-300 series stretched upper deck can accommodate 80 people, and the complete double deck 700-plus-seater Boeing 747 is not far away. However, the next generation 1000-seater aircraft is, for the moment, some way off.

On international flights, routes, schedules and fare structures are agreed bilaterally between governments via agencies such as the Civil Aviation Authority (CAA) in the UK and the Federal Aviation Authority (FAA) in the United States. Airline applications to operate services at certain fare levels are negotiated between governments for approval. On certain routes services are pooled; each national airline operates an equal number of flights and revenue received is divided equally between the two. There are also a number of international aviation bodies of which ICAO and IATA are examples. ICAO — the International Civil Aviation Organization — is a UN agency comprising government representatives who set technical standards relating to equipment, maintenance, flight procedures and safety, etc. Although most comply with ICAO, notable exceptions are the USA and USSR who set their own standards, and do not conform.

In 1944 the rights of international carriers were agreed when 52 nations signed the Chicago convention that established the five freedoms of the air. Freedom one lays down the right to overfly a country; two, the right to make a technical stop in a country for refuelling and servicing only; through to the fifth and now sixth freedoms, which allow, for example, an international airline to uplift and set down fare-paying passengers between two other nations, e.g. Pan American operating New York - London - Brussels has negotiated rights to board passengers in London and disembark them in Brussels.

IATA — the International Air Transport Association — is a trade federation comprising airline representatives from member companies who set commercial standards relating to passenger facilities, in-flight services, seating (including seat pitch), catering (including meal composition), baggage allowance, etc., and even 'free items' such as drinks or gifts. Members are expected to conform to recommendations. Many airlines find it an advantage to remain outside IATA but most of the big operators prefer the security of membership.

* * *

The time is now 1845 with Skyship One lying 200 n.m. off the coast of Newfoundland, level at 350. The last HF radio position report was passed to Gander at 52N 50W giving an estimate for Dotty of 1903, and Skyship One was instructed to call Gander on VHF 126.9 MHz approaching Dotty. On the flight deck the crew have finished their meal and are preparing charts for the next phase of the journey, while in the cabin, movies shown after the meal service are just ending and a few passengers are moving about stretching their legs. The last weather copied from New York at 1810 on HF frequency 8868 kHz indicated JFK forecasting low cloud and rain. At 35,000 feet over the western Atlantic the present wind is 270°T at 50 knots with an OAT of -62°C. Ground speed is 437 knots.

F/O R/T: - Gander, Skyship One, level three five zero, estimating Dotty one nine zero three.

Gander R/T: - Skyship One, good afternoon, squawk five six two four.

(5624 is selected on the transponder).

Gander R/T: - Skyship One, you are radar identified, cleared to Kennedy via Dotty, North American route one six zero, maintain level three five zero. Omit position reports.

North American route 160 is confirmed from the Western supplement as High Level and Jet 581 to Conay, Fredericton,

Bangor, Kennebunk (and on to New York), and is checked against the flight logs.

Over Newfoundland with approximately 2½ hours to go, the Captain speaks to the passengers giving a position report and informs them that the arrival time is still estimated on schedule at 4.35 p.m. local, and that the New York weather is still much as expected. At 1910 the weather is again copied, this time on HF frequency 5652 kHz, and indicates JFK as:- cloud 900 feet broken, 2,200 feet overcast, visibility 3 miles, rain, fog, temperature 40°F, dew point 38°F, wind 180°T at 13 knots. Skyship One is now instructed to call Gander on frequency 133.9 MHZ.

> *F/O R/T:* - Gander, Skyship One heavy, good afternoon, level three five zero, requesting level three seven zero and any direct routeings.

(Heavy is now added to the callsign to indicate a big jet).

Skyship One is asked to standby while the computer checks the routeings to New York. Direct routeings are often requested at this stage, and with the aircraft still too heavy for level 390, opposite direction flight levels, such as 370 in this case, are frequently available, with most traffic travelling in the same direction. Presently clearance is obtained for level three seven zero and climb initiated. Over the Gulf of Saint Lawrence, thirty minutes later, clearance is also given to proceed from present position direct to Norwich, 100 n.m. north of New York, a total direct distance of over 600 n.m. Norwich is in INS way-point 9, and as 0-9 is keyed on the INS the aircraft turns to fly the great circle track direct, saving time and fuel. Along route beacons are, of course, still tuned in to check progress, but aircraft fly mostly abeam positions on the direct paths. Flying now at level 370 the aircraft is under Moncton control on frequency 133.1 MHz, and crossing the New Brunswick coast at Prince Edward Island is passed to Moncton frequency 132.15 MHz.

At 2014 abeam Fredericton the aircraft is only five minutes from the US border and control changes to Boston on 134.8 MHz. Skyship One is instructed by Boston to squawk 2333 and the direct routeing to Norwich at flight level 370 is confirmed, with a Saybrook 3 arrival at New York. The *en route* Boston frequency is quiet as position reports are not required in the USA and there is little R/T. Selcal is already selected on VHF box 3, which is tuned to the Arinc New York frequency of 129.9 MHz, and Selcal chimes are heard indicating that contact is being attempted. (Arinc is a US communications company that relays commercial and private messages throughout its network in the USA. The system can also be connected to ground telephone lines.) Replying on 129.9 MHz,

Skyship One is passed a Company message by Arinc informing of 30 minutes landing delays at JFK because of weather. There is insufficient fuel for 30 minutes delay at New York plus a diversion to Montreal if JFK closes, but a quick look at the Boston weather copied earlier indicates Boston above limits and the Captain now nominates Boston as the diversion airport. The Flight Engineer checks the Boston diversion fuel from Kennedy as 14.9 tons, which, with an estimated 20.0 tons remaining overhead, leaves 5.1 tons for holding delays. Holding fuel is calculated at a rate of 9.0 tons per hour, which allows 34 minutes holding for the 5.1 tons available — just right for the expected delays.

Abeam Bangor in Maine now at 2031, still level 370, Skyship One is passed to Boston control frequency 134.95 MHz, and preparations are made for descent, approach and landing. The standard arrival route Saybrook 3 allocated earlier is examined from the charts (Fig. 11.10) and shows the routeing to be along the 238° radial out of Norwich VOR to Calverton VOR, then the 263° radial out of Calverton to Deer Park VOR via Micke, an imaginary point (9 DME from Deer Park), which is the cleared limit of the Saybrook 3 arrival. The hold, or stack, at Micke is shown as a left-hand race-track pattern inbound on the 082° radial from Deer Park VOR, which is

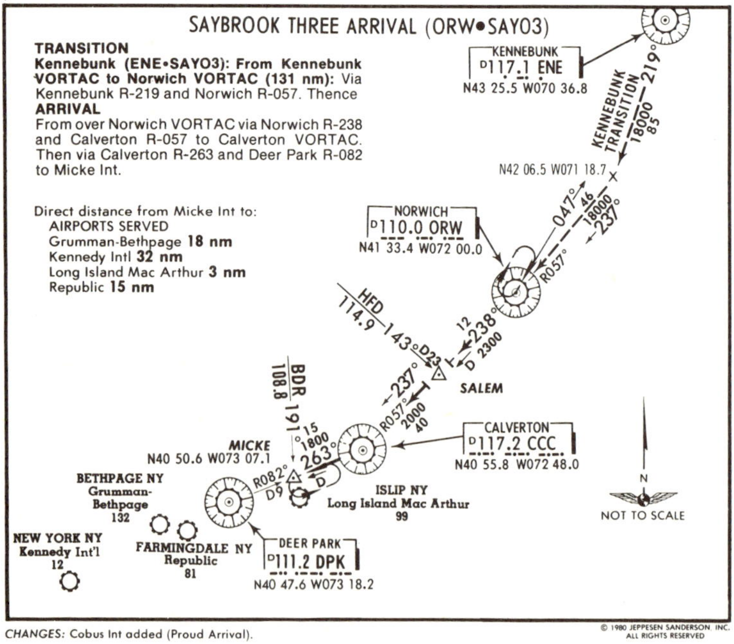

CHANGES: Cobus Int added (Proud Arrival).

Fig. 11.10 Saybrook three arrival chart. (Courtesy Jeppeson Sanderson Inc.).

Elev	Var	INS	PAX	CARGO
12	12°W	RAMP	N 40 38.9 W073 46.9	N 40 39.6 W073 48.0

(J.F.KENNEDY INTL) **NEW YORK**
AERODROME

KJFK

START UP	KENNEDY	Ground	Tower	NEW YORK	VOT	ATIS	D1
Refer **B3**	**Clearance**	**121.9** 348.6	**119.1**	**Departure**	**115.1**	**115.1**	**8 JUN 81**
	135.05 348.6	**121.65**	258.3	**121.1** 269.0			

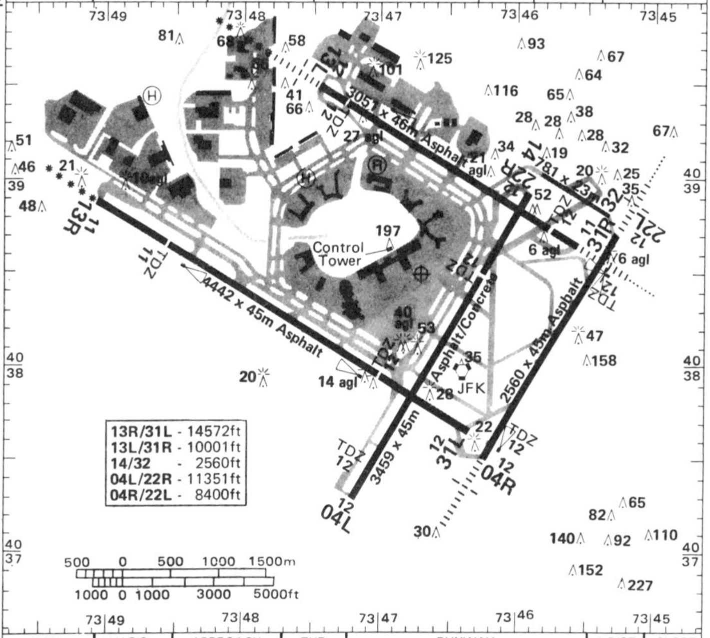

R/W	VASIS	APPROACH	THR	RUNWAY	L.DIST	SLOPE
04L(031°T)	Nil	Nil	Gr + S	HRL CL YeCZ	Full	Nil
22R(211°T)	S		Gr		2538m	
04R(031°T)	Nil	ALSF-1	Gr	TDZ HRL CL YeCZ	Full	Nil
22L(211°T)		MALSR				
13L(121°T)	S	LDIN ALSF-1	Gr	TDZ HRL CL YeCZ	2743m	0.01D
31R(301°T)	Nil	MALSR		HRL CL YeCZ	2733m	0.01U
13R(121°T)	S	LDIN	Gr	HRL CL YeCZ	3649m	0.01U
31L(301°T)	Nil	Nil	Gr + S		3428m	0.01D
14 (125°T)	Nil	Nil	Gr + S	MRL	720m	Nil
32 (305°T)			Gr		Full	

OTHER LIGHTING: Emergency, obstruction, taxiway, A/D beacon.

RUNWAY AVAILABILITY
14/32 - propeller aircraft only upto 5670kg.
RUNWAY SURFACES
04L/22R, 13L/31R & last 914m(3000ft) of 04R/22L grooved.
TAXIWAY WIDTHS
All 23m(75ft)

Rev: Frequencies

1. Aircraft landing on 04L/R should exercise extreme care due to turbulence caused by jet aircraft taking-off on 31L.
2. When 04R/22L is wet, braking action reports to be requested from ATC.

D1 AERODROME (JOHN F. KENNEDY INTL) **NEW YORK**

Fig. 11.11 New York (JFK) runway chart. (Courtesy AERAD).

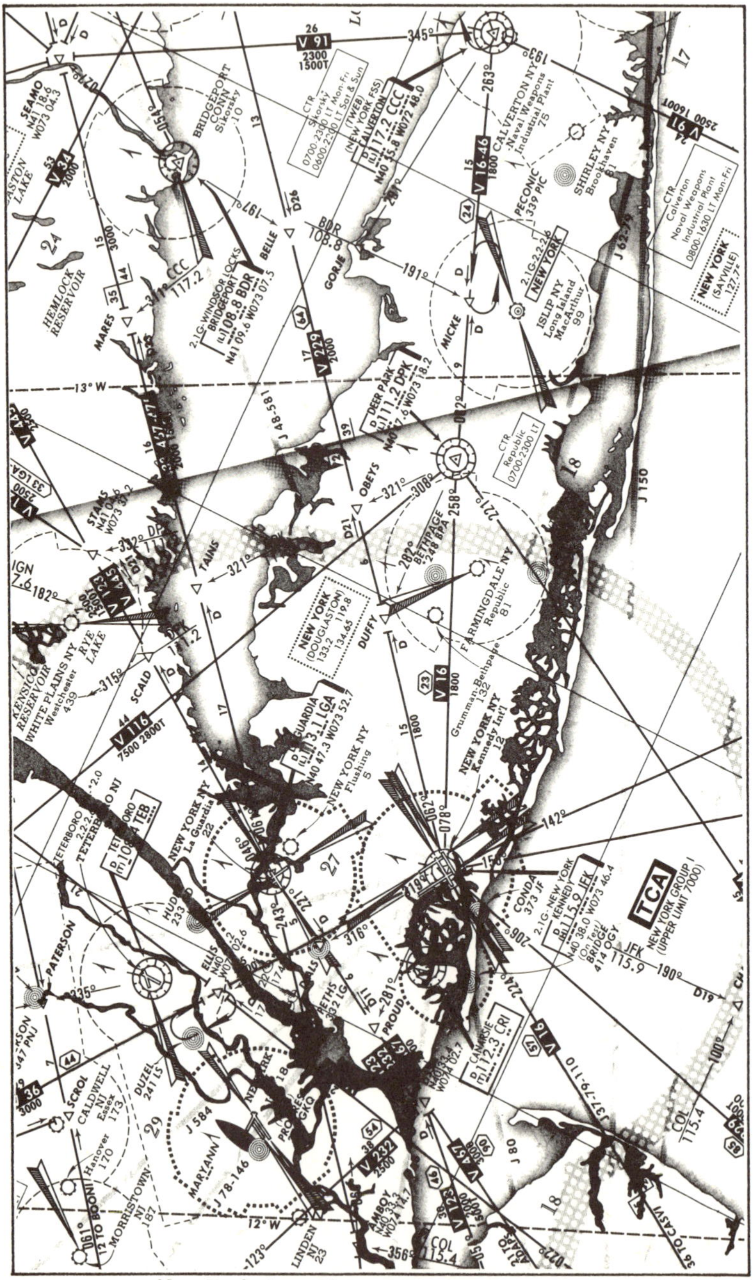

New York Area Chart (Courtesy Jeppesen Sanderson Inc.)

262°M track direction. When absorbing delays, aircraft, unfortunately, cannot stop in mid-flight like traffic at a red light, and have no choice but to circle; but precise patterns are flown over a point rather than aircraft just arbitrarily going round in circles.

The last New York weather forecast indicated a wind of 180°T at 13 knots (192°M — variation 12°W) so a landing can be expected into wind on runway 22 left or 22 right (JFK has two groups of parallel runways lying NE/SW — i.e. 04L and R, 22R and L — and NW/SE — i.e. 31L and R, 13R and L — like the disjointed lines of a noughts and crosses game. Fig. 11.16). The track from Deer Park to the 22 runways is approximately 36 n.m. so a height of around 10,000 feet at Deer Park VOR is considered about right. As a rough guide the distance needed to loose height is calculated as the height loss required divided by 300, i.e. 10,000 — 300 = 33 n.m. To arrive at 10,000 feet over Deer Park at 250 knots, the maximum speed allowed below that height in the United States, it is necessary to extract from a table, showing height to be lost against aircraft weight and wind component, the distance from Deer Park to commence descent. In this case the height to be lost is 27,000 feet (from level 370 down to 10,000), the aircraft weight is around 230 tons, and the average head wind component in the descent is estimated as 25 knots. The distance found is 75 n.m., which in fact is about the track distance from Norwich VOR to Deer Park VOR (actual 76 n.m) so descent can be commenced over Norwich, about 100 n.m. north of New York.

However, in spite of having calculated the best descent point, ATC in the United States is notorious for descending aircraft early and the crew prepare for descent some time before Norwich. Approach and landing procedures for the 22 runways are also studied and from the landing chart (Fig. 11.12) are noted safety heights, approach and missed approach procedures, and other details such as ILS frequency and inbound track, locator outer marker position and height to cross above the marker on final approach (as a check of the ILS glide path), runway threshold displaced or normal, approach lighting and runway length. The stack at Micke is also checked and the entry procedure established. The crew at this stage, of course, do not know for certain that the 22 runways are in use and some last minute planning may be required, but they do know they have a Saybrook 3 arrival with a possible holding delay at Micke, and the approach is assumed to be radar vectors from Deer Park to the ILS for 22 left or right.

Just north of Boston with about 15 minutes to go to Norwich, estimating overhead at 2100, control changes to frequency 134.47 MHz. The Boston weather is copied from the ATIS on Boston VOR frequency 112.7 MHz, with the voice facility selected, and is

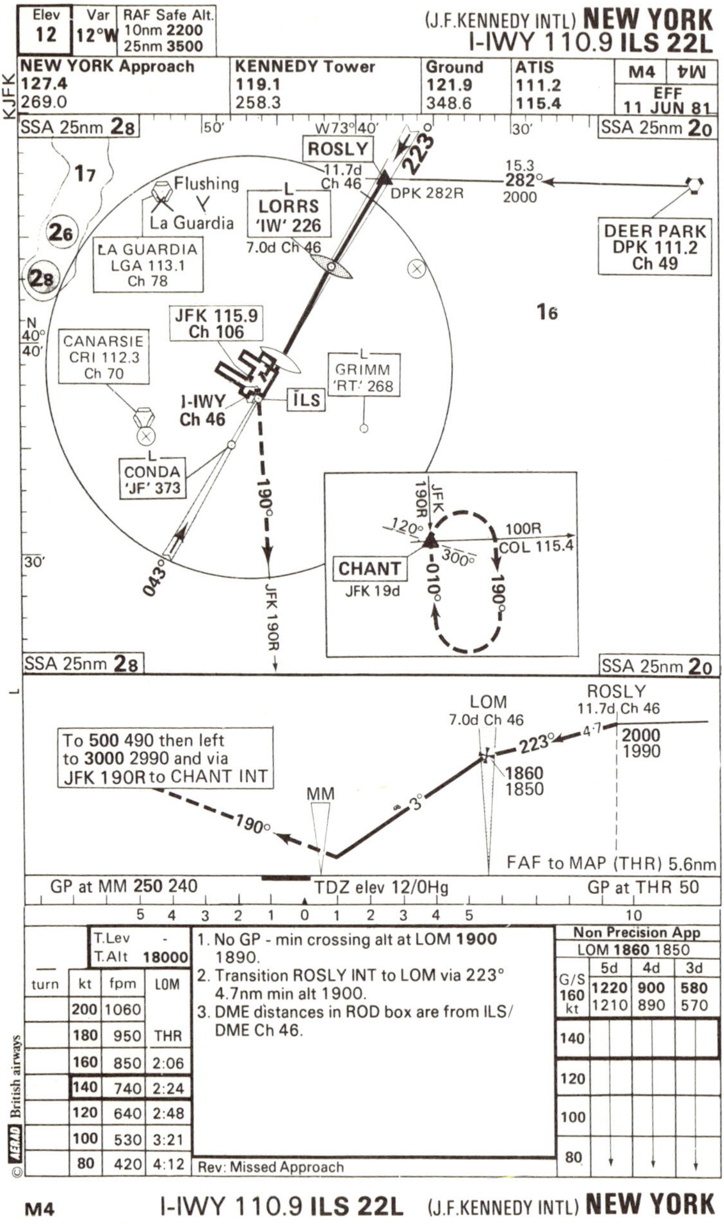

Fig. 11.12 Landing chart. (Courtesy AERAD).

confirmed as suitable for diversion. With the aircraft close to descent now the Captain calls for the 'before descent check'. Once again the Flight Engineer reads from the check list as the other two respond to calls. The landing weight is estimated at 223 tons and a minimum speed for crossing the runway threshold with full flap, the reference speed — V ref, is extracted from a table against weight. The indicated V ref speed of 132 knots is noted on a landing data card together with the minimum speeds for flap 20° (V ref + 10 — 142 kts), flap 5° (V ref + 40 — 172 knots) and zero flap (V ref + 80 — 212 knots) all based on the reference speed. The engine power required in the event of a missed approach is calculated for the present conditions as 1.41 EPR and is also noted on the landing data card. Other information included is safety heights, locator outer marker height, missed approach height and landing limits (i.e. the minimum cloud-break height and minimum acceptable visibility), in this case 320 feet above sea level (from the pressure altimeter) and 4000 feet visibility for a hand-flown ILS and 200 feet above the ground (from the radio altimeter) and 2000 feet visibility for an autocoupled approach with a manual landing, the visibility being the important factor. Autoland is approved on runway 22L but is not calibrated for actual weather conditions less than those required for an autocoupled approach.

Since the JFK weather is not too good the Captain intends to fly the aircraft automatically down the ILS to a minimum height, without visual contact, of 200 feet above the ground on the radio altimeter, at which point — the decision height — go-around must be initiated if the approach lights are not in view. If contact is established at decision height approach can be continued, the autopilot disengaged and the aircraft landed manually, or automatically if desired in the case of 22 left. If the autopilots are disengaged at any time during the approach before ground contact is established the decision height reverts to the hand-held limit of 320 feet on the pressure altimeter. All the speeds shown are bugged on the ASI together with 80 knots, the speed at which reverse thrust is normally reduced. The heights of 200 feet and 320 feet are also bugged on the radio and pressure altimeters respectively. The Flight Engineer sets the pressurization for landing, humidifiers are switched off, instrument switching is checked, seats and safety harnesses are checked locked and secure, landing data and limits are checked and set, safety heights are confirmed, and the Captain briefs the crew on the approach procedures. In this case an autocoupled approach to 22 left or right is intended and ILS and beacon selection, locator outer marker height, landing limits and missed approach procedures are all discussed. The Flight Engineer announces the 'before descent' check complete.

Boston R/T: - Skyship One heavy, cleared pilot's discretion descend to and maintain flight level two two zero, to cross Norwich level two two zero.

Because of traffic, descent has been given much earlier than hoped but the crew have no option other than to comply. Since initial descent is rapid it is estimated that 40 n.m. out from Norwich is sufficient to commence descent to lose 15,000 feet and be level by Norwich.

Descent and holding

F/O R/T: - Boston, Skyship One heavy is out of level three seven zero for level two two zero.

Using the vertical speed knob the Captain winds the aircraft rate of descent to 3000 feet per minute and on leaving 370 calls for descent power. The Flight Engineer firmly closes the throttles and the aircraft literally glides in the descent with minimum power set. The descent speed is initially M 0.82, but as the aircraft descends into the more dense atmosphere the ASI becomes the primary speed instrument and the rate of descent is adjusted to about 5000 feet per minute until the IAS increases to 340 knots. Speed mode is engaged and descent is continued at 340 knots with the rate of descent easing off. As the aircraft descends, ice on the inside of the cold metal edges of the flight deck windows, formed in the cruise with the humidifiers switched on, begins to melt as the outside air warms and drips uncomfortably on the crew.

Within range of the JFK ATIS transmitted from the Deer Park VOR the Flight Engineer copies the details of information 'bravo' — cloud 400 feet scattered, 900 feet broken, 1700 feet overcast, visibility three statute miles, rain, fog, wind 195°M at 15 knots, temperature 40°F, dew point 38°F, altimeter 29.50 (inches of mercury), landing runway ILS 22 left. The Flight Engineer also contacts Company on VHF box 2 and advises them of ETA.

Capt. PA: - Ladies and gentlemen, good afternoon, this is the Captain. We have now commenced our descent into New York but have been informed of a thirty minute landing delay because of weather, and we now estimate landing in about fifty minutes' time. I'll keep you informed of any improvement. The New York weather is low cloud and rain with fresh winds, and the present temperature is forty degrees Farenheit, five degrees centigrade.

Leaving Norwich at level 220, being navigated by the INS along the Saybrook three arrival route, Skyship One is passed to New York on 133.3 MHz. The pace hots up now with the volume of radio chatter increasing markedly. Skyship One is further cleared to 12,000 feet on the altimeter setting of 29.50. The transition level everywhere in the United States is flight level 180, and since cleared below this height, 29.50 is set and checked on both altimeters. The instruments now read the aircraft height above mean sea level in the New York area. Further clearance is received to descend to 10,000 feet, to be level by Micke, and to take up the hold at Micke. At the moment the flight is still automatic, with crew members monitoring progress and mapping a mental picture of aircraft position from the instruments. Micke is still some way off and a little power is applied to slow the rate of descent. The INS is tracking the aircraft a bit right of track now and heading mode is engaged and the aircraft turned back on track using the heading knob and guidance from the instruments. Passing through 11,000 feet at 340 knots over Calverton VOR the Captain turns the aircraft onto the 263° radial to Micke.

All: - One thousand to go.

Control passes now to New York Approach on frequency 125.7 MHz and the R/T becomes a constant flow of instructions and replies.

F/O R/T: - New York Approach, Skyship One heavy just levelling ten thousand.
NY Approach R/T: - Skyship One heavy, maintain ten thousand. Once round the hold at Micke then direct Deer Park.

Obviously the thirty minute holding delay has been cleared and Skyship One requires only one four minute holding pattern for flow sequencing. The aircraft is now only 13 n.m. from Micke and 'altitude green' is called levelling at 10,000 feet as the speed begins to drop off. At this height the aircraft is in the 6000 to 14,000 feet bracket and within this layer in the United States is expected to hold at 210 knots. The minimum speed bugged for zero flap is 212 knots, so permission is requested and granted from Approach to hold at Micke at 230 knots to keep the aircraft clean and save fuel. The holding speeds required in the United States (from ground level to 6000 feet — 200 knots; 6000 feet to 14,000 feet — 210 knots, and above 14,000 feet — 230 knots), and ICAO regulations in the same layers (210 knots, 220 knots and 240 knots), are designed to maintain aircraft within the holding areas, but permission is often granted for higher speeds to allow aircraft to fly clean (without flap.)

The throttles are fully closed now to help reduce speed as the aircraft is aerodynamically well designed and even level at 10,000 feet it takes about 10 n.m. to lose the 110 knots from 340 knots to 230 knots for holding. Deer Park is selected on both the VORs and the inbound holding track at Micke of 262° is set on the course indicator. Micke is 9 DME from Deer Park and, as the distance indicates on the instruments with the speed at 230 knots and level 10,000 feet, the aircraft is turned by the Captain using the heading control onto the outbound leg of 082° for one minute, and the heading adjusted to allow for drift. With each turn in the hold approximately one minute and the outbound leg flown for about one minute, adjusted for wind, a four minute holding pattern is achieved. The entry into the Micke hold for Skyship One is easy as the aircraft simply turns left onto the outbound leg, but entering the same hold from other directions requires different entry procedures. On occasions aircraft are instructed, unexpectedly, to hold at beacons at short notice and some quick thinking is required to work out the holding pattern, the sector in which the aircraft lies, and the procedure. (During training, all students, at some time, turn the wrong way!)

At the end of one minute on the outbound leg the Captain turns the aircraft left towards Micke and adjusts the heading to establish on the inbound leg with the beam bar central in the instruments. The aircraft has been cleared for one hold and at Micke continues towards Deer Park.

> *N.Y. Approach R/T:* - Skyship One heavy, recleared eight thousand. Leave Deer Park heading two five zero. When level eight thousand, speed two ten knots.

Skyship One complies, and as the aircraft is slowed flap 1° and the approach check are called for. The First Officer checks the speed, selects flap 1° and checks that the flaps are set — 'flap one set' — while the Flight Engineer proceeds with the approach check. Cabin signs are switched on, boost pumps are switched on, cross-feed valves are checked set for landing, ignition is set to flight start, speed brake is armed to deploy automatically at touchdown, auto brakes are set to minimum, and altimeter settings are checked. The approach is not considered finished until 5° flap is set, but all other items are complete.

> *Capt. PA:* — Ladies and gentlemen, the long holding delays have not materialized and we are now commencing the approach. We expect to land in about fifteen minutes' time. From here we fly west down Long Island then turn left for final approach to land towards the south-west.

At exactly 2114 the aircraft crosses Deer Park and turns left onto a heading of 250°, height level at 8000 feet, speed 210 knots. The New

York local time is 4.14 p.m. and it is still light but extensive cloud cover lies below with the ground not yet visible. Skyship One is now under radar control and is cleared to Approach frequency 132.4 MHz.

Approach and landing

Today's approach under radar is controlled almost entirely from the ground with ATC instructing aircraft on headings, heights and speeds. At modern radar control centres controllers have indications on their screens of aircraft track, height and ground speed, but are unable to tell heading, airspeed or rates of climb and descent, and where required have to ask the pilots for the information. Aircraft are flown in 'S' turns (and even circles) or slowed prematurely, all under radar control, for sequencing onto final approach at the appropriate separation. Radar headings finally position flights at an angle of approximately 30° to 40° to the runway centre line about 10 n.m. from the threshold, the capture of the ILS and final approach being flown, either automatically or manually, under control of the pilots.

Where radar control is not available at less busy airports descent clearances are still required, but descent, approach and landing procedures, including heights and speeds, are all planned and controlled by the crew. In such cases ATC clearances normally involve descent to overhead a radio beacon at the airport at a certain height, say 4000 feet, and it is up to the pilots to arrive overhead at the correct height and speed with the appropriate checks complete ready for an approach. With poor weather a full instrument procedure is carried out which normally involves positioning outbound and descending to a certain level by flying away from the airport in the opposite direction to the runway centre line inbound track, completing a procedure turn to fly the aircraft inbound towards the airport, then along the runway centre line track descending on final aproach to landing. Let-down charts for all runways with instrument approaches are carried on board and indicate procedures and descent profiles in plan and elevation, with much relevant detail added.

Airports with full radar coverage also indicate full let-down procedures to be flown in the event of equipment unserviceability. Precise tracks have to be flown by adjusting heading to allow for drift, stop watches are used to time legs and turns, and speeds, heights and descent profiles have to be accurately flown. In bad weather conditions a great deal of concentration is required to fly the aircraft accurately while buffeting in cloud. If the approach facility is an ILS, runway centre line and glide path guidance are available to the touch-down point, and the aircraft can be flown accurately either by hand with reference to the instruments or automatically by the autopilots flying down the radio beams of the ILS to a manual or automatic landing. Some ILS approaches at certain airports are monitored by precision approach radar in bad weather and pilots warned of any significant deviation from the ILS — for example Hong Kong, where runway 31 ILS approaches between the islands. VOR and NDB beacon approaches are only cloud-break procedures and are hand-flown right down to landing, with visual reference being required on short finals. A VOR approach is less accurate than an ILS approach and requires higher minimum cloud base limits, while an NDB approach requires still higher limits, being the least accurate of all.

When approaching a quiet airport without radar in good weather the Captain may request a visual approach and landing, which involves positioning the aircraft relative to the runway by looking out of the window (just like a light aircraft) while flying height, heading and speed accurately from the instruments. Positioning visually for landing saves time and avoids tedious procedures but, of course, if an ILS or other facility is available it is used for guidance on final approach. In fact, where aircraft do position visually it is imperative to cross check from available aids that the aircraft is lined up on the correct runway if parallel runways are in use, and, of course, at the right airport. Such airports as, for example, Sharjah and Dubai in the Persian Gulf lie close to one another with runways of the same direction, and it has not been unknown for pilots to land at the wrong one!

Weather for visual approaches must comply with visual meteorological conditions (VMC) which stipulate that the pilot must be in sight of the ground, visibility must be at least 5 n.m., and horizontal and vertical separation from cloud at least 1 n.m. and 1000 feet respectively. Where weather conditions are worse than these, instrument meteorological conditions (IMC) apply and the pilot is expected to remain on instruments until in visual contact with the runway at a certain height above the landing point. In VMC aircraft comply with visual flight rules (VFR), which basically state that pilots are responsible for their own separation between flights and are not subject to air traffic control clearances or instructions. In IMC, at night, and in controlled airspace, even in clear weather, aircraft must comply with instrument flight rules (IFR), and certain specific regulations are mandatory. All the big jets operate mostly within controlled airspace and are therefore subject to IFR, which stipulates that an ATC flight plan must be submitted with details of the flight, that ATC clearances and instructions must be adhered to, that certain appropriate radio equipment must be carried, and that pilots must be suitably licensed, i.e. hold a current instrument rating.

During training on all aircraft from single engine props to the big jets, pilots practise take-offs and landings with a procedure known as 'circuits and bumps'. A circuit comprises a rectangular pattern involving take-off into wind followed at a safe height by a climbing turn to track at right angles to the runway and levelling off at a certain height (1500 feet for the big jets). Once level the aircraft is turned downwind to fly parallel to the runway until past the threshold, followed by another 90° turn to descend at right angles towards the runway and ending with a last turn onto the runway centre line, continuing descent to landing. When slowing on the landing roll quick preparation is made for another take-off and, without stopping, take-off power is set and the aircraft leaps into the air again. Flying circuits involves all the basic flying procedures from take-off, climb, turn, level flight, descent approach and landing.

Each circuit only takes about ten minutes and is demanding flying with high work load, but the experience can be put to good use when positioning visually for approach to land. Instead of overflying the airport and completing a full circuit the pilot visually positions the aircraft on the downwind or base leg (either left or right depending on the direction approaching the airport), by judging distance and height from the runway using skills gained during circuit practice, and then flies the remainder of the circuit down to landing. If the airport lies directly ahead the pilot can line up for a straight-in approach.

When flying visually to land, normally some kind of radio facility is available for guidance on final approach, but only ILS (if installed) indicates glide path. Without radio facilities, maintaining runway centre line is not difficult as the runway can be seen ahead and approach path lighting shows clearly. However, judging the correct descent profile is more difficult and a number of aids have been designed for visual guidance. The visual approach slope indicator (VASI) consists of two coloured bars of lights placed one above the other, situated at both sides of the runway by the touch-down point. When the aircraft is too high the bars show yellow, when too low red, and when on the glide slope the top shows yellow and bottom red. In the Far East, Australia and New Zealand a 'T' bar system of lights is in use whereby if a 'T' is seen the aircraft is low, an inverted letter 'T' indicate the aircraft is high, and if only the cross-bar is seen the glide path is being flown.

The latest system is the precision approach path indicator (PAPI), which consists of a horizontal row of four coloured lights placed on either side of the touch-down point. The aircraft is on the glide slope

Approach to Hong Kong. Aircraft established on the IGS flying towards the Checker Board. The airport is on the right.

Commencing the turn to finals.

two reds and two yellows are showing; too high if more than two yellows are showing, and too low if more than two reds are showing. Many ILS runways also have visual guidance systems installed as a check against the ILS glide slope or for use during visual approaches when equipment is withdrawn for maintenance, or has failed.

Where such visual facilities are not installed, or where an approach is being made in cloud without ILS using a VOR or NDB, glide slope guidance is not available and some simple mental calculations are required as an aid. ILS and visual approach systems normally indicate glide paths at an angle of 3° to the horizontal which in fact allows for a descent on the approach of 300 feet per nautical mile. If distance to touch-down is known from DME (INS is often not accurate enough after long journeys), the height required at any point on the approach is simply the distance to the threshold times 300, i.e. 5 n.m. to touch-down — height required 1500 feet; 3 n.m. — height required 900 feet, etc. During such approaches the vertical speed indicator is an important instrument and the rule is simply to multiply the ground speed indicated by the INS by 5 to obtain the rate of descent required, e.g. 140 knots ground speed, rate of descent required = 140 × 5 = 700. In this case descending at 700 feet per minute on the vertical speed indicator will maintain a 3° glide slope.

Where a locator outer marker is used for the approach the aircraft is flown over the beacon at the height published in the let-down charts, for example 1600 feet. The time taken from the locator outer marker to touch-down is also published on the charts against aircraft ground speed. If the time given is, say, two minutes, then clearly 800 feet per minute rate of descent is required from the beacon to a sea level runway. Over the locator outer marker the stop watch is started as a countdown to the threshold, which allows a further check to be made of the approach path at the one-minute-to-go point. If at one-minute-to-go the altimeter height is 750 feet, then obviously a rate of descent of 750 feet per minute is required for short final approach to the touch-down point.

Visual approaches, i.e. using eyeball and judgement as opposed to ILS are not uncommon, even at larger airports, if only for the simple reason that equipment has to be withdrawn for maintenance, or perhaps a landing is required on a little used runway without ILS because of wind conditions. On such visual approaches, however, it is very unusual not be have some kind of visual glide path aid or some simple means of calculation. Judging height and distance in a big jet just by looking at the angle to the runway is very difficult, although, of course, on the rare occasion the pilot may have no choice. One example may be where a new runway is built and has, as yet, no aids whatsoever. Some visual approaches may also be required for noise

At 500 feet in the turn.

Passing Checker Board at 500 feet.

abatement or traffic sequencing purposes, or because of terrain problems on the approach. The ILS approach to runway 13L at JFK, for example, is seldom used as aircraft are required to overfly the Manhattan and Queens areas of New York, and disruption of traffic flow to La Guardia just 9 n.m. north of JFK results. Instead approaches are made at right angles to the runway, over the Canarsie VOR beacon to the west of JFK, followed by a 90° turn onto the runway, commenced below 1000 feet at about 3 n.m. from touchdown. Heavy-jet pilots are advised to be stabilized on the approach by 800 feet at the latest, and large turns at low level close to the threshold are not easy, especially in cross-wind conditions. On 13L, however, strobe lights on the ground and VASIs are available for guidance, and useful visual aids include the Aqueduct racetrack some 2 n.m. from the threshold and the white façade of the International Hotel on short finals on the extended runway centre line.

Among the Caribbean Islands, Antigua, for example, has reasonable facilites with a VOR and two NDB radio beacons, but the use of such aids on the approach is restricted because of terrain. On the approach to an easterly landing on runway 07 VASIs are available for final approach, but the only reference available to the pilot downwind is the visual sighting of the runway, and when turning base leg even that disappears from view behind a hill. The approach then continues over the town of St. John's and onto short finals for landing between the hills. Not the easiest of airports on a wet and windy night!

Hong Kong is another airport with terrain problems, a straight-in approach towards the south-east not being possible because of mountains lying to the north. Instead when landing on runway 13 an approach is made initially on a system similar to an ILS but offset at 47° to the runway, and referred to as an instrument guidance system (IGS). The IGS signals are rejected by the pilots at about 700 feet and the aircraft is flown towards a large red and white checker board before commencing a sharp 40° turn onto the runway at about 500 feet. The IGS has a paired DME with a built-in delay, which gives distance to the landing threshold, and on the ground strobe lights mark out the turn with offset VASIs indicating glide path. (In Hong Kong only aviation danger and guidance lights are permitted to flash. All commercial advertising lights must be steady).

* * *

Skyship One is now heading 250°, level 8000 feet, speed 210 knots and under radar control with New York Approach on 132.4 MHz.

At 400 feet in the turn.

Lining up on finals at 300 feet.

The aircraft is about 20 n.m. from JFK heading south-west towards the airport, and the Captain selects the 22L ILS frequency 110.9 MHz (ident I-IWY) on his side with the 22L inbound track of 223° selected in the course window to set the correct presentation of the ILS beam bar for the runway centre line. The First Officer has the JFK VOR frequency 115.9 MHz already selected on his side with the DME indicating 20 n.m. (and also, for convenience, 223° in the course window). Both aids are identified by their morse coding. The ILS has an associated DME which is cross-checked with the VOR DME distance and compared against the INS distance as an INS accuracy check.

New York Approach R/T: - Skyship One heavy, turn left heading one three zero for sequencing. Traffic one o'clock moving right to left, height unknown.
F/O R/T: - Skyship One heavy, left heading one three zero. We're looking for the traffic.

Numerous light aircraft are in the vicinity, which radar control have on their screens and whose positions are indicated to the big jets. Radio communications are one continuous flow with one controller handling many aircraft.

New York Approach R/T: - Skyship One heavy, turn left heading zero eight zero.
F/O R/T: - Skyship One heavy, left heading zero eight zero. Where are we going now?

The aircraft has been instructed to head towards the east, up Long Island away from New York.

New York Approach R/T: - Skyship One heavy, I'm going to have to take you all the way round again for sequencing. Turn left now three six zero, speed back to one eighty knots, recleared four thousand.
F/O R/T: - Roger Skyship One heavy, left three six zero, speed one eighty knots, down to four thousand.

The Captain closes the throttles and calls for flap 5°. The Co-pilot selects the flaps, and as the speed comes back to 180 knots the Captain lowers the nose for descent using the vertical speed knob. With flaps set at 5° the Flight Engineer calls the approach check complete.

Passing 7000 feet the aircraft enters cloud and engine anti-icing is switched on and engine power increased slightly to maintain sufficient supply for hot air. Skyship One buffets in the cloud with occasional heavy rain lashing the windshield, and the odd lightning flash can be seen.

> *New York Approach R/T:* - Skyship One heavy, turn left heading three three zero.

Turning left onto 330° 'one to go' is called passing 5000 feet, and 'altitude green' as the aircraft levels out at 4000, still in cloud and experiencing light turbulence.

> *New York Approach R/T:* - Skyship One heavy, turn left two four zero, descend now to two thousand, cleared ILS two two left, minimum speed one sixty till the outer marker. Call the tower one nineteen one when established.

Distance to touch-down is now 15 n.m. with the aircraft placed on a suitable heading to intercept the 22L ILS. Flap 10° is called and checked set. Skyship One establishes in the descent at 180 knots, with some power still applied for anti-icing purposes. At the Captain's request the First Officer selects and identifies the 22L ILS frequency on his side, and both locator beacon frequencies of 226 kHz are also selected and identified. With all ILS and locator facilities selected and checked the Captain switches to land mode on the autopilot panel and engages autopilot 'B'. The two autopilots are now engaged in preparation for an autocoupled ILS approach. Autopilot 'A' continues to fly the aircraft until the dual green light illuminates at 1500 feet and both autopilots take control. As Skyship One levels at 2000 feet 'altitude green' is called again with the aircraft 8 n.m. from touch-down, heading 240°, and still no sight of the ground. Power is applied to about 1.1 EPR and with the speed settled at 180 knots the autothrottle is engaged. The crew are glued to the instruments, carefully monitoring the automatic progress. 'Nav. green'. The navigation green lights illuminate as the autopilot captures the ILS localiser, the beam bars move from off the stops on the right of the instrument and swing towards the centre, followed by the aircraft automatically turning sharply left to track down the runway centre line. The Captain maintains hands lightly on the controls throughout the automatic approach. Almost immediately the glide slope indicators are seen to move down from the top of the instruments as the glide path is intercepted from below. 'Glide slope active' is called.

> *Captain:* - Flap twenty please.

Flap 20° is checked set, and as the Captain adjusts the autothrottle bug to 170 knots the throttles automatically ease back on the power. At glide slope capture 'glide slope green' is called and the aircraft nose dips in descent as the autopilots fly the glide slope.

F/O R/T: - Tower, Skyship One heavy established ILS two two left.
New York Tower R/T: - Skyship One heavy, continue approach. Call me at the marker.

Autopilot 'B' unexpectedly drops to off with the warning light flashing amber, but a quick check indicates systems normal and the autopilot is successfully re-engaged. The Captain trickles the autothrottle speed back to 160 knots and calls for the gear.

Captain: - Gear down, landing check.

The Co-pilot selects the gear down and a green light indicating all is well is checked by each in turn. A last check of the hydraulics by the Flight Engineer confirms systems satisfactory and the landing check is called complete. Over the locator outer marker the ADF needles swing round and marker lights flash blue with intermittent tone. Stop-watches are started (time from LOM to threshold is noted as 2 min 24 sec) and altimeter heights are compared with the published locator outer marker height of 1860 feet as a check against the glide slope.

F/O R/T: - Skyship One heavy, outer marker inbound.
New York Tower R/T: - Skyship One heavy, cleared to land. Runway is wet — braking action reported as good.

Where runway surfaces are contaminated with rain, snow or ice, braking action is reported in terms of good, medium or poor. Automatic brakes can be set to minimum, medium or maximum, depending on conditions. With minimum set Skyship One needs a stopping distance of 6700 feet at a landing speed of 137 knots. Runway 22L is 8400 feet long. Most runway lengths at big airports are, in fact, about 10,000 - 12,000 feet long (2 n.m./2.3 s.m./3.5 km), but notable exceptions are Doha in the Persian Gulf and Harare in Zimbabwe with runway lengths of 15,000 feet (2.5 n.m./2.8 s.m./4.6 km). The normal minimum Boeing 747 landing distance is 7000 feet.

Skyship One is now 5 n.m. from touch-down passing 1500 feet in thick cloud, speed still 160 knots. The wind indicates 180°T at 20 knots on the INS as the aircraft crabs along the runway centre line heading 218°, 5° to the left. 'Dual green' is called as the light illuminates indicating both autopilots in control. The landing flap setting (flap 30°) is called by the Captain, selected by the First Officer, and checked set by the crew, and the autothrottle speed is reduced to 137 knots (5 knots above V ref.). The aircraft pitches markedly nose downwards with full flap selection and the autothrottles ease the power to reduce speed.

On occasions on final approach an aircraft may be required to go around again for a number of different reasons. It may be that the runway is blocked owing to a landing or taking-off aircraft not clearing sufficiently quickly, the flight is catching up on the aircraft ahead, or the visibility falls below limits. Depending on the circumstances the crew may be instructed by ATC to carry out a missed approach procedure, or the Captain may dislike the situation and initiate action himself. The Captain calls 'go-around, flap twenty' and pulls back on the control column (or palms the automatic go-around switches) while at the same time advancing the thrust levers. The Co-pilot selects the flaps to 20° and the Flight Engineer finely adjusts the throttles to go-around power. Gear is selected up by the Co-pilot on the Captain's command, when a positive rate of climb is established. Instrument routeings and heights for missed approach procedures are published on let-down charts and are memorized by pilots before approach, although at most big airports radar control is available for repositioning to final approach.

Approaching 1000 feet, 'the one thousand feet' check confirms gear and full flap are set, no warning flags, and the missed approach height of 3000 feet set in the altitude select window. Suddenly the aircraft breaks cloud and the runway is seen clearly 3 n.m. ahead. Light rain is falling, and the 5° drift can be detected from the approach angle, the runway appearing slightly to the right. With the runway in view the Captain disconnects the autopilots and autothrottle and continues the approach hand-flying down to a manual landing. The momentum of the B747 results in a fairly stable flight path being flown, but any deviation has to be positively corrected by ailerons and elevators. The rudder pedals, of course, remain central, only being used in asymmetric flight and for guidance along the runway on take-off and landing.

At this stage of the approach the aircraft has to be kept under firm control and coarse flying control and thrust lever movements may be required, especially in gusty winds. However, even in stable conditions, continuous small corrections have to be made to controls and thrust levers to adjust for small changes in wind speed and direction. During the final stages of the approach the Captain not only monitors runway centre line and glide slope indications on the flight director bars, and beam bar and pointers, but attitude, speed, heading, altitude, rate of descent, engine power setting and time to touch-down, all form part of the scan, and the eyes flow in a continuous movement from instrument to instrument. Power requirements at this stage are around 1.05 EPR but any power increase results in a pitch-up attitude (with engine pods slung below the wings), which has to be corrected by down elevator to maintain the ILS glide slope and likewise, power reductions have to be corrected by up elevator. Passing 500 feet, with 45 seconds to go, the wind direction is transmitted by the tower as 190°M at 15 knots giving 13 knots head wind and 3° right drift. The indicated airspeed shows 138 knots (V ref + ½ head wind component), ground speed 125 knots, the heading 220°, with the flight director bars crossed in the centre indicating the Captain flying the ILS accurately. The Co-pilot and Flight Engineer carefully monitor progress.

On short finals, when aircraft are close to the ground, certain wind conditions can cause wind shear, which can be a hazard to flight. Wind shear is a rapid change of wind direction or speed affecting the airflow over the wings, and can occur without warning. On approach, drift and wind component are monitored carefully by the First Officer for any indication of change, drift being read from the INS and wind component by comparing true airspeed (TAS) from the gauge with ground speed from the INS. A higher TAS indicates head wind, lower, tail wind. Wind shear detection equipment is under test at the moment, whereby a laser beam measures reflected light from dust particles 300m ahead of an aircraft and, using indicated ground speed, computes expected TAS in advance. A dropping TAS indicates wind shear. Heavy rain may also be a hazard owing to the drag of the rain and water droplets causing wind surface roughness, and thunderstorms in the vicinity of airports can cause severe downdraughts. In most cases pilots simply delay take-off or approach until storm clouds have departed. Wake turbulence can be a further problem on approach, with aircraft landing or taking-off ahead creating wing tip vortices — large horizontal whirlpools of air with tangential velocities which when generated by a C5A or Boeing 747 can be up to 90 knots. The whirlpool cores can be 25 to 50 feet across, and rapidly expand rearwards from each wing tip until they overlap and dissipate. Landing times between Jumbos and smaller aircraft are normally increased but wake turbulence is not as great a hazard to large jets.

As Skyship One passes 420 feet the First Officer calls 'One hundred to go', and at 320 feet, 'Decision height'. The Captain replies 'Continuing'. The Captain's eyes now dart from instruments to runway in judging the final stages of the approach, while the First Officer monitors the radio alitmeter to call precise heights above the runway.

F/O: - One hundred feet. Fifty feet (the threshold flashes by). Thirty feet.

At 30 feet the Captain flares the aircraft by raising the nose a few degrees to arrest the rate of descent and at the same time closes the thrust levers. (Air squeezed between the aircraft under-surface and the ground acts in cushioning the touch-down and is known as ground effect). Although height is called from the radio altimeter by the Co-pilot, manual touch-down still requires a good degree of height judgement and the Captain's eyes look well down the runway to improve the angle. (As when stopping by a roadside in a vehicle, distance out can be judged more easily by looking ahead rather than directly at the edge.) The aircraft is still crabbing because of drift with the nose pointing left of the runway centre line, and just before touch-down right rudder is applied to ease the nose straight. With the yawing movement in straightening the aircraft the left wing swings into wind and the extra lift produced raises the wing which has to be counteracted by left down aileron. On the point of touch-down the Captain's right hand is on the closed thrust levers, the left hand is on the control column, easing back to judge the flare, and at the same time applying left aileron to hold the wings level, with the right foot feeding in rudder to keep straight. The sequence of controlled movements lasts only a few seconds and correct timing becomes important in strong cross winds. Skyship One lands smoothly at 2129 G.M.T., and on touch-down the spoilers automatically deploy. Passengers, of course, appreciate a good landing, but in rainy or windy conditions it may, on the grounds of safety, be more prudent to land firmly to make good contact with the wheels on the runway through the wet and to prevent drifting off the runway in strong cross winds.

The touch-down speed is around 130 knots (150 mph/257 km/hr) and at a weight of approximately 220 tons the aircraft requires quite an amount of stopping. On touch-down the wheels spin up with a puff of smoke and immediately the autobrakes (which are protected by anti-skid devices) operate to slow the aircraft. The Captain gently eases the column forward to lower the nose wheel while keeping the wings level with aileron, and at the same time selects reverse thrust with the right hand, pausing until all deflector doors are positioned before applying power. As the engines roar in response the Captain keeps the aircraft straight down the runway using rudder. Reverse power is effective at high speeds, and as the thrust is felt to bite, slowing the aircraft, the autobrakes modulate in response. At 80 knots reverse thrust is cancelled, and the First Officer calls the speed from 80 every ten knots down to 10. As the aircraft slows autobraking is set to off and manual braking applied using the toe brakes, body gear steering is selected, and the Captain's left hand transfers to the tiller for steering. Skyship One leaves the runway by a suitable exit and is instructed to remain on tower frequency.

New York Tower R/T: - Skyship One heavy, proceed right zulu, left golf, hold short two two right.

The parallel runway is also active. Most big airports have lettered or numbered taxiways and crews follow instructions from charts. Occasionally 'follow me' trucks are available to guide aircraft to the gate. The Captain calls for the 'after landing check' and flaps and speed brakes are selected in, and flight equipment switched off. The Flight Engineer starts the auxiliary power unit (APU) on the taxi. Approaching 22R Skyship One is cleared to cross and instructed to call ground on 121.9 MHz when clear.

F/O R/T: - Ground, Skyship One heavy, clear two two right.
New York Ground R/T: - Skyship One heavy, first right on the outer taxiway to the gate.

At the designated gate number the aircraft slows to park nose in to the terminal building as the First Officer calls the speed (8 knots) from the INS. A line on the ground extending from the terminal indicates the parking bay centre line and the Captain turns the aircraft to track the nose wheel along the line. The Flight Engineer instructs the cabin staff on the PA to set the doors to manual and the 'after-landing check' is complete. Parking has to be accurate to position aircraft correctly for the passenger movable walkways, and normally some kind of stand guidance system indicates precise positioning to the Captain. Stand guidance systems vary, but parking bay centre lines are normally indicated on the face of the terminal by a small vertical bar required to be kept in line, or by two vertical green indicators. The latter is known as the azimuth guidance for nose-in stands (AGNIS) and indicates that an aircraft is off the stand centre line by illuminating the appropriate bar red.

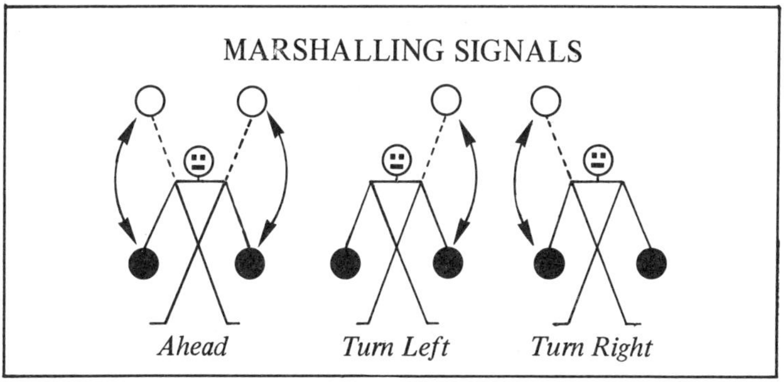

Fig. 11.13.

Stopping is sometimes indicated on a board to the side by lining a marker against the aircraft type (parallax aircraft parking aid — PAPA) and occasionally by a yellow and black striped bar placed across the bay at windshield height. Where guidance sysems are not available, or where parking is away from the terminal building, aircraft are marshalled to the stopping position. Marshallers used red coloured bats by day and lit batons at night. (Fig 11.13). Both arms are moved from in front to above the head to indicate to the pilots to move straight ahead, and the respective arm to indicate turns. Stop is indicated by crossing arms above the head. At certain airports, such as Washington (Dulles) in the United States, giant mobile lounges position by aircraft and transport passengers up to 100 at a time to the terminal building.

As Skyship One approaches the stopping position the First Officer continues to call the speed as the Captain monitors the stand guidance indicators. At the exact spot the aircraft is stopped, the parking brake set, and the Captain calls for the 'leaving aircraft check'. The nose wheel chocks are confirmed set by ground and the chock time of 2136 (one minute late on schedule) confirmed. With power transferred to the APU, engines are shut down, seat belt signs are extinguished, and flight equipment is switched off in turn. Before the INS are shut down comparisons are made between the actual aircraft position and position as indicated by the INS. Numbers 1, 2 and 3 INS indicate differences of 3, 5 and 13 n.m. respectively — not bad after 3000 n.m. Flight instruments are set for the 'before start check' for the next crew, paper work is collected and filed, and charts are folded and put away. A ground power unit is connected as soon as possible, and with the auxiliary power unit shut down the 'leaving aircraft check' is complete.

Thirty minutes after chocks under, the crew are deemed to be off duty. It is just after 5.00 p.m. New York time, 10.00 p.m. London time. Tomorrow the crew may be rostered to fly the next leg of the westbound round the world service, or perhaps the eastbound service back to London, or if they are New York based, will return home for a few days break before the next flight. While they rest, tens of thousands of crew members will be flying thousands of aircraft around the clock to all points of the globe. Only occasionally do the aircraft get a break!

Appendix
The Big Jets

Boeing 747.

Boeing 747—100, 200 and 300 series

Wing span: 195.62 ft/59.64m.
Length: 231.76 ft/70.66m.
Height: 63.40 ft/19.33m.

Fuel capacity: 100 series, 183,570 litres/40,380 Imp. gallons/48,650 US gallons; 200 series, 198,380 litres/43,640 Imp. gallons/52,580 US gallons; 300 series, 198,380 litres/43,460 Imp. gallons/52,580 US gallons.

Maximum take-off weight: 100 series, 332,900 kg/734,000 lbs; 200 series, 371,900 kg/820,000 lbs; 300 series, 377,850 kg/833,000 pounds.

Maximum seating capacity: 100 series, 500; 200 series, 550; 300 series, 660.

Range with full load: 100 series, 3940 n.m./4020 s.m./6460 km; 200 series, 5800 n.m./6680 s.m./10,900 km; 300 series, 5680 n.m./6540 s.m./10,500 km.

Full tanks range/remaining capacity load: 100 series, 7280 n.m./8380 s.m./13,480 km; 200 series, 7300 n.m./8400 s.m./13,850 km; 300 series, 7300 n.m./8400 s.m./13,850 km.

*Normal cruise altitudes:*28,000ft. to 39,000ft. *Normal cruise speed:* Mach 0.80 to 0.85.

Engines (4): 100 series, Pratt and Whitney JT9D-7 (46,950 lbs thrust) or General Electric CF6-45A2 (46,500 lbs thrust); 200 series, Rolls-Royce RB211-524D (53,110 lbs thrust) or General Electric C56-50E2 (52,500 lbs thrust) or Pratt and Whitney JT9D-7Q (53,000 lbs thrust); 300 series, Pratt and Whitney JT9D-7R4G2 (54,750 lbs thrust).

Lockheed Tristar – L1011.

Lockheed Tristar L-1011-500 (Long range)

Wing span: 164.3 ft/50.09 m. *Length:* 164.1 ft/50.04 m. *Height:* 55.3 ft/18.86m.

Fuel capacity: 119,776 litres/26,347 Imp. gallons/31,740 US gallons.

Maximum take-off weight: 224,900 kgs/496,000 lbs.

Maximum seating capacity: 330.

Range with full load: 4600 n.m./5300 s.m./8520 km.

Full tanks range/remaining capacity load: 6700 n.m./7715 s.m./12,410 km.

Normal cruise altitudes: 28,000 ft to 39,000 ft.

Normal cruise speeds: Mach 0.80 to 0.85.

Engines (3): Rolls-Royce RB211-525 (50,000 lbs thrust).

McDonnel Douglas DC10.

McDonnel Douglas DC10-30 and 40 (Long range)

Wing span: 165.28 ft/49.17 m. *Length:* 181.38 ft/55.30 m. *Height:* 58.0 ft/17.68 m.

Fuel capacity: 138,740 litres/30,380 Imp. gallons/36,600 US gallons.

Maximum take-off weight: 251,700 kg/553,700 lbs.

Maximum seating capacity: 380.

Range with full load: 5380 n.m./6200 s.m./9670 km.

Full tanks range/remaining capacity load: 6425 n.m./7400 s.m./11,800 km.

Normal cruise altitudes: 28,000 ft to 39,000 ft.

Normal cruise speeds: Mach 0.80 to 0.85.

Engines (3): (DC10-30) General Electric CF6-50C1 (52,500 lbs thrust); (DC10-40) Pratt and Whitney JT9D-59A (53,000 lbs thrust).

Abbreviations

a.c.	Alternating current
ADF	Automatic direction finder
ADI	Attitude director indicator
AGNIS	Azimuth guidance for nose-in stands
AH	Artificial horizon
AIDS	Airborne integrated data system
Airep	Airborne weather reports
ALT	Altimeter
AM	Amplitude modulated
A/P	Autopilot
APU	Auxiliary power unit
ASI	Airspeed indicator
A/T	Autothrottle
ATA	Actual time of arrival
ATC	Air traffic control
ATCC	Air traffic control centre
ATIS	Automatic terminal information service
ATPL	Airline transport pilots' licence (UK)
ATR	Airline transport rating (Canada and USA)
C	Compass
°C	Degrees centigrade or degrees compass
CAA	Civil Aviation Authority (UK)
CAB	Civil Aeronautics Board (USA)
CAT	Clear air turbulence
CAVOK	Ceiling and visibility OK
Cb	Cumulonimbus cloud
Ch	VOR associated DME channel number (e.g. Ch 56)
CO2	Carbon dioxide
C of G	Centre of gravity
C of P	Centre of pressure
CP	Critical point or equal time point
CPL	Commercial pilots' licence
CRT	Cathode ray tube
CSD	Constant speed drive
CTA	Control area
CTR or CTZ	Control zone
Cu	Cumulus cloud
CVR	Cockpit voice recorder
CW	Carrier wave
d.c.	Direct current
DDM	Dispatch deviation manual
DG	Directional gyro
DME	Distance measuring equipment
EGT	Exhaust gas temperature
EPNdB	Equivalent perceived noise decibels
E/O	Engineer Officer
EPR	Engine pressure ratio
EPRL	Engine pressure ratio limit
ETA	Estimated time of arrival
ETD	Estimated time of departure
ETP	Equal time point or critical point
°F	Degrees Fahrenheit
FAA	Federal Aviation Authority (USA)
FCU	Fuel control unit
F/D	Flight director
FDR	Flight data recorder

FF	Fuel flow
FFCC	Forward facing crew cockpit
FIR	Flight information region
FL	Flight level
FM	Frequency modulation
FMA	Flight mode annunciator
FMS	Flight management system
F/O	First Officer
g	Force of gravity
°G	Degrees grid
GCA	Ground control approach
G/E	Ground engineer
G.M.T.	Greenwich mean time
G/P	Glide path (same as glide slope)
Gradu	Gradually
G/S	Glide slope
HDG	Heading
HF	High frequency
HP	High pressure or horse power
HSI	Horizontal situation indicator
HUD	Head up display
Hz	Hertz
IAS	Indicated airspeed
I.A.T.	International atomic time
IATA	International Air Transport Association
ICAO	International Civil Aviation Organisation
IFR	Instrument flight rules
IGS	Instrument guidance system
ILS	Instrument landing system
IMC	Instrument meteorological conditions
Imp	Imperial
IR	Instrument rating
INS	Inertial navigation system
ISA	International standard atmosphere
ITCZ	Inter-tropical convergence zone
kg	Kilogram
kg/hr	Kilograms per hour
kHz	KiloHertz
km	Kilometre
km/hr	Kilometres per hour
lbs	Pounds
LE	Leading edge
LF	Low frequency
L.M.T.	Local mean time
Loc	Localiser
LOM	Locator outer marker
LP	Low pressure
LSB	Lower side band
LW	Long wave
LWT	Landing weight
m	Minute or metre
°M	Degrees magnetic
mb	Millibar
Met	Meteorology
MF	Medium frequency
MHz	MegaHertz
mph	Miles per hour
MSL	Mean sea level
MW	Medium wave
N	Compressor spool
NAV	Navigation
NC	Compass north
NDB	Non-directional beacon
NG	Grid north
n.m.	Nautical mile
NM	Magnetic north
Ns	Nimbostratus
NT	True north
Nosig	No significant change
OAT	Outside air temperature
OCA	Ocean control area
OM	Outer marker
OSV	Ocean station vessel
PA	Public address
PAPA	Parallax aircraft parking aid
PAPI	Precision approach position indicator
PAR	Precision approach radar
Pax	Passengers
PCU	Power control unit
PNdB	Perceived noise decibels
PNR	Point of no return
PPI	Plan position indicator

Prob	Probability
p.s.i.	Pounds per square inch
QFE	Airport elevation altimeter pressure setting
QNH	Mean sea level altimeter pressure setting
°R	Degrees radial
RA	Radio altimeter
RMI	Radio magnetic indicator
ROC	Rate of climb
ROD	Rate of descent
rpm	Revolutions per minute
R/T	Radiotelephony
RTOW	Regulated take-off weight
RVR	Runway visual range
s	Second
SAT	Saturated air temperature
Sc	Stratocumulus cloud
S/E/O	Senior Engineer Officer
S/F/O	Senior First Officer
Sigmet	Significant meteorological broadcast
SHP	Shaft horse power
SHF	Super high frequency
SID	Standard instrument departure
s.m.	Statute mile
S/O	Second Officer
SP	Special performance
SSA	Sector safe altitude
SSB	Single side band
SSR	Secondary surveillance radar
SST	Super sonic transport
STAR	Standard terminal arrival routes
SW	Short wave
°T	Degrees true
TA	Transition altitude
TAF	Terminal area forecast
TAS	True airspeed
TAT	Total air temperature
TE	Trailing edge
TL	Transition level
TMA or TCA	Terminal control area
TOW	Take-off weight
TRU	Transformer rectifier unit
T & S	Turn and slip
TURB	Turbulence
UHF	Ultra high frequency
UIR	Upper flight information region
U/S	Unserviceable
USB	Upper side band
UTA	Upper control area
VASI	Visual approach slope indicator
V1	Go or no-go decision speed
VR	Rotation speed
V2	Take-off safety speed
VFR	Visual flight rules
VHF	Very high frequency
VLF	Very low frequency
VMC	Visual meteorological conditions
VMCG	Minimum speed for control on the ground
Volmet	Plain language weather broadcast
VOR	Very high frequency omni-directional radio range
VOT	VOR test facility
V/S	Vertical speed
VSI	Vertical speed indicator
W/V	Wind velocity
Wx	Weather
ZFW	Zero fuel weight
Zulu time	Military G.M.T.

Index